*Textiles
and
Industrial
Transition
in Japan*

DENNIS L. MCNAMARA

Textiles and Industrial Transition in Japan

Cornell University Press · ITHACA AND LONDON

First published 1995 by Cornell University Press.

Library of Congress Cataloging-in-Publication Data

McNamara, Dennis L.
    Textiles and industrial transition in Japan / Dennis L. McNamara.
        p.   cm.
    Includes bibliographical references and index.
    ISBN 0–8014–3100–X (acid-free paper)
    1. Textile industry—Japan.   I. Title.
    HD9866.J32M38   1995
    338.4'7677'00952—dc20   94–43122

⊛ The paper in this book meets the minimum requirements of the American National Standard for Information Sciences— Permanence of Paper for Printed Library Materials, ANSI Z39.48-1984.

To Mary Coffey,
Ellen O'Connell,
and Marie Shannon,
for their wisdom, patience, and care

Contents

Tables

ABBREVIATIONS

ACTWU   Amalgamated Clothing and Textile Workers Union
ASEAN   Association of Southeast Asian Nations
DSP   Democratic Socialist Party
EPA   Economic Planning Agency
FAS   Foreign Agricultural Service
GAO   General Accounting Office
GTC   General Trading Company (Sōgo Shōsha)
ICAC   International Cotton Advisory Council
ILGWU   International Ladies Garment Workers Union
ILO   International Labour Organisation
ITGLWF   International Textile, Garment, and Leather Workers' Federation
ITMF   International Textile Manufacturers Federation
JCC   Joint Consultation Committees (Rōshi Kyōgi Kaigi)
JCFA   Japan Chemical Fibers Association (Nihon Kagaku Sen'i Kyōkai)
JPC   Japan Productivity Center (Nihon Seisansei Honbu)
JSA   Japan Spinners' Association (Nihon Bōseki Kyōkai)
JTN   *Japan Textile News*, published monthly by Osaka Senken Ltd
LDP   Liberal Democratic Party
MITI   Ministry of International Trade and Industry (Tsushō Sangyōsho)

NBG	*Nihon Bōseki Geppō* (Monthly Report of the Japan Spinners' Association)
OECD	Organization for Economic Cooperation & Development
TWARO	Textile Workers: Asian-Pacific Regional Organization
USDA	United States Department of Agriculture

PREFACE

A nation's textile industry can tell us much about its economic history. Textiles usually stand at the forefront of early industrial efforts, with labor-intensive production of commodity goods for the local market. Mills provide jobs, serve local demand for clothing and other textile products, and offer a productive investment outlet for local capital, bringing multiple benefits to state and society in the process of industrialization. The Japanese government and, later, private entrepreneurs imported British spinning machines in the late nineteenth century to produce cotton goods for the domestic market. Japan's mills soon turned their growing energies abroad, exporting cotton textiles to China, Korea, and other Asian nations. This turn to international markets followed a path familiar among industrializing nations, and indeed among textile producers—but with one exception. In the absence of local cultivation of raw cotton, Japanese mills had to import cotton from India and the United States. Early Japanese shippers and trading houses grew and prospered with the cotton trade, importing cotton and exporting textiles.

The textile industry brought together millers, traders, bankers, and merchants, of course, and soon they were joined by machinery-makers who produced spinning and weaving machines locally. Embedded in the economy and society since the late nineteenth century, and soon concentrated in a small number of huge companies, the Japanese spinning industry found a place among world leaders by the 1920s. A variety of heavy and light industries in Japan had diversified industrialization well before the Pacific War, and although the textile industry remained a leading employer and exporter, problems of overproduction had already forced mills into temporary production cartels. Much of the nation's mill capacity in Osaka and Tokyo was destroyed in World War II. National reconstruction helped the millers rebuild with new machinery and quickly restore production for markets at home and abroad. Armed with new machinery and new investment, local mills were outpacing the domestic demand for textiles in the 1950s. In the next decade, they found themselves losing traditional export markets as Asian countries established their own textile industries. By the 1970s, low-cost Asian competitors had begun to erode Japan's share in the lucrative markets of the United States and Europe. And as Asian competitors in the 1980s turned their export energies to the Japanese market, local mills began losing market share at home. Decline since the 1950s has prompted a remarkable effort to revive and sustain the industry through a series of restructuring programs.

Unlike many of their competitors in the West, Japan's larger textile mills have weathered the restructuring of their "declining" industry. Their history offers a profile of transitions in an advanced, industrialized economy and speaks to us of adjustment, corporatism, and collective behavior. Survival in any declining industry is remarkable, but the survival of Japan's larger textile mills owes less to the market and more to corporatist strategies of adjustment worked out by capital, labor, and the state. Since rising costs of labor, land, and energy in Japan may soon bring comparable pressure to bear on the automobile, heavy machinery, electronics, and other industries, the experiences of the textile industry may contain lessons in industrial adjustment that could prove valuable to all. Also this chronicle of change may offer some insight into Japanese society, for the restructuring of the textile industry occurred in a collective context that extends limited, specific, or "special" interests into the more comprehensive, general interests of the community.

Restructuring did not significantly alter the leadership of the textile industry established by the "big ten" spinners and the "big nine" synthetic-fiber makers or "moguls" (ōte kaisha) shortly after World War II. Among these, rankings in sales and market share vary little from year to year. Dominance in the market permits the moguls to exercise leadership when negotiating with labor and the government. A hierarchy of large firms weathering the restructuring process might suggest simple collusion, but closer scrutiny reveals a curious blend of unity and divergence, of consensus and dissent even in this oldest of Japan's modern industries. Survival of a few of the leading mills can be traced to dissent and divergence from a carefully orchestrated consensus, a flexibility that merits special attention in this closely inter-connected industry. The rapid rise of a few "mavericks" draws attention to a competitive, combative dynamic in this transition period in the textile industry, for even as moguls were scrapping capacity, mavericks were adding it, and as moguls were reducing textile production in natural fibers, mavericks were expanding such production. How can we explain the persistence of the mogul hierarchy, yet the emergence of newcomers? Strategies of adjustment helped to moderate decline and maintain hierarchy in the industry, without precluding the rise of a few mavericks.

Industries in capitalist economies must constantly adapt and restructure to sustain their role in international trade. One finds market opportunities for both labor-intensive production of commodity textiles and capital-intensive production of higher-quality items. Such opportunities have stimulated investment in both developing and advanced economies to maintain and expand local textile production. Japan's experience sheds light on the process of industrial restructuring and on the organization of an international network of textile manufacture and trade, but it also reveals patterns of mediation that encourage competitive interests to find common cause in shared paths of adjustment. A study of change in the industry draws us into embedded patterns of competition and cooperation that channel enterprise within Japanese society. For instance, the "textile business" (sen'i gyōkai) provides a vantage point for tracing the negotiation of a public "interest" with multiple private "interests" in a world riven by the competing priorities of state and capital, industry and individual firm, labor and management, mill and merchant.

A corporatist structuring of interests helped to moderate decline and maintain stability without discouraging competition. Changes at mills, supported by traders and apparel houses, indicate moderate, gradual decline, and bargaining among labor, capital, and the state offer evidence of corporatist structures and programs that eased the transition. That neocorporatist patterns of change proved decisive in the restructuring of textiles is significant both within and beyond Japanese society. This book thus sheds light on domestic patterns of cooperation and competition among capital, state, and labor. Attention to modes of collective action unveils deeply rooted and widely routinized patterns of conflict management in Japanese society. Beyond Japan, the book helps refine theories of neocorporatism and collective behavior useful for comparative political economy. The project begins and ends at the mills, with a systematic study of adjustment in one of the world's leading textile industries.

I started with many interviews, a good deal of scrutiny of company records, and long discussions. Intensive interviews at the mogul and maverick firms taught me initially of the general direction of change, and only later did I learn of individual cases of divergence and dissent. What began with a sectoral study of corporatist adjustment at the mills concluded with an analysis of collective behavior in an entire industry. What began with spinning companies ended up with a society of firms and their workers, of related industries, government offices, and offshore affiliates. Tensions between mogul and maverick firms turned my attention to a complex realm of competition and cooperation within this society of industry. A prominent executive in Osaka initially alerted me to the moguls' problems with maverick spinners: "We now have two competitors—the LDCs [less developed countries] and the Nagoya spinners." His comment provided the first inkling of seams in what I had thought was a seamless garment of industry relations.

When I returned to the industry associations with new questions about solidarity among the firms, I was quickly drawn into a turbulent business world rent by divisions and patched together by compromises. Officials candidly explained how the competition between moguls and mavericks initially developed and how it continues to play out in the powerful Japan Spinners' Association (JSA). The story of adjustment was no longer a case study of firms adjusting to markets, but a far more complex drama of people, firms, and markets. I gradually pieced together a chronicle of moguls and mavericks vying for technology, products, and markets and angling for government favor at home and abroad. Meetings with labor leaders, traders, and the executives of apparel firms helped round out the picture of an "industry" in transition. I learned of the changing role of the state in conversations with officials at the Ministry of International Trade and Industry (MITI), textile executives, and labor officials busy at the Diet. Interviews with industry executives at Japanese affiliates, and with local industry leaders in Jakarta and Bangkok, turned my attention to an international network of production and trade centered in Osaka and Tokyo.

***

Research took me to Tokyo, Osaka, and Nagoya, where I combined interviews with a review of archives at industry associations, labor unions, and government offices. Company reports provided information on scale, products, sales, and employment, but the project came alive with the help of leaders in the business world of textiles in Japan and abroad who alerted me to networks, divisions, and compromises. A study of corporatism, collective action, and industrial change also involves an array of intellectual mentors, of course, and my citations catalog the scholars who have helped shape my ideas. Among them Ezra Vogel, Andy Gordon, and T. J. Pempel deserve special thanks. A grant from the National Science Foundation supported a year of research in Japan, where the Institute of International Relations at Sophia University provided a base for research and wonderful colleagues. Such support is gratefully acknowledged. I hope insights culled from this journey through time and place may offer direction for industry and society in subsequent transitions.

DENNIS L. MCNAMARA

Washington, D.C.

1   Adjustment in Decline

Albert O. Hirschman defines "interest" as "the disciplined under-standing of what it takes to advance one's power, influence and wealth." Max Weber writes of the rational calculation of interest as a distinguishing mark of capitalist society. J.A.W. Gunn and Albert Hirschman trace the notion of interest to statecraft in sixteenth- and seventeenth-century Europe, as monarchs hoped to control individual "passions" such as avarice and cunning in the emerging market economies with commitment to a "public" interest.[1] Identification of interest with profit or economic advantage grew out of this dialogue and the continuing expansion of market economies. Ideas of both interest or profit (*rieki*), and of interests or "advantages and disad-vantages" (*rigai*) took center stage in Japanese capitalism in the late nineteenth century and never relinquished it. In this book I look to three interest groups involved in the restructuring of Japan's textile industry: state, capital, and labor.

Scholars of Japanese capitalism emphasize the communal character of private enterprise and entrepreneurship, reinforced by a prominent state role in early industrialization.[2] Ezra Vogel has recently distin-guished Japanese business ideology as "adaptive communitarianism": "From the beginning of Meiji [i.e., 1868] the large companies in key sectors had the view that, while it was permissible to make money and

1. Albert O. Hirschman, *The Passions and the Interests* (Princeton: Princeton Uni-versity Press, 1977), 38; J.A.W. Gunn, "Interests Will Not Lie: A Seventeenth Century Political Maxim," *Journal of the History of Ideas* 29 (October–December 1968): 551–64.
2. Johannes Hirschmeier and Yui Tsunehiko, *The Development of Japanese Busi-ness* (Cambridge: Harvard University Press, 1975); Arthur E. Tiedemann, "Big Busi-ness and Politics in Prewar Japan," in *Dilemmas of Growth in Prewar Japan*, ed. James William Morley (Princeton: Princeton University Press, 1971), 267–316.

while in a certain sense they had the rights of private ownership, they were bounded by and responsible to the government and national interest."[3] Byron Marshall cites a consensus in prewar business ideologies that individual interests and private gain "should be subordinated to the preservation of harmony within an organic society."[4] Arnold J. Heidenheimer and Frank C. Langdon write similarly of an organic view of society among postwar entrepreneurs, noting how entrepreneurs preferred to bargain through intermediaries in the political realm, rather than dominate the political process overtly.[5] It is this embedded character of economic relations that draws our attention.[6]

An organic emphasis on general or communal over individual gain has deeply influenced structural adjustment in Japan. The market economist Martin Bronfenbrenner was shocked to find anxiety among Japanese business and government leaders over "excessive competition" (*katō kyōsō*).[7] More recently, Daniel Okimoto returned to the theme with a general definition of such competition: "a zero-sum situation in which an excess number of producers possess supply capacities that far exceed demand."[8] But Bronfenbrenner had initially

3. "Japan: Adaptive Communitarianism," in *Ideology and National Competitiveness: An Analysis of Nine Countries*, ed. George C. Lodge and Ezra F. Vogel (Cambridge: Harvard Business School Press, 1987), 148.

4. *Capitalism and Nationalism in Prewar Japan* (Stanford: Stanford University Press, 1967), 115.

5. *Business Associations and the Financing of Political Parties* (The Hague: Martinus Nijhoff, 1968), 148. See also two articles by Ishida Takeshi, "The Development of Interest Groups and the Pattern of Political Modernization in Japan," in *Political Development in Modern Japan*, ed. Robert E. Ward (Princeton: Princeton University Press, 1968), 293–336, and "Interest Groups under a Semipermanent Government Party: The Case of Japan," *Annals of the American Academy of Politics and Social Science* 413 (May 1974): 1–10.

6. For a description of socially embedded features of economic life, see Mark Granovetter, "Economic Action and Social Structure: The Problem of Embeddedness," *American Journal of Sociology* 91 (November 1985): 481–510; see also Granovetter and Richard Swedberg, "Introduction," in *The Sociology of Economic Life*, ed. Granovetter and Swedberg (Boulder, Colo.: Westview Press, 1992), 1–28.

7. Martin Bronfenbrenner, " 'Excessive Competition' in Japanese Business," *Monumenta Nipponica* 21, nos. 1–2 (1966): 114–24. Shindo Takejiro traces excessive competition to the low entry costs in the industry: "The fact that spinning enterprises can be undertaken on a small scale is one of the causes of excessive competition in the industry." *Labor in the Japanese Cotton Industry* (Tokyo: Japan Society for the Promotion of Science, 1961), 223.

8. *Between MITI and the Market: Japanese Industrial Policy for High Technology* (Stanford: Stanford University Press, 1989), 38.

argued that excessive competition in Japan connoted a situation "driving a good company [*yūryō kaisha*] out of some important line of its business," "forcing the default of a company's loans to a 'city bank,' " or "forcing good Japanese firms to bear risks of unfavorable development." The problem continues to draw attention. A former MITI official, Matsumoto Koji, contrasted the appropriate competition of firms and "assurance of suitable profits" with the destructive competition of firms seeking for "maximum expansion of total sales or shares."[9]

The term "appropriate competition" (*tekitō kyōsō*) appears prominently in documents of MITI and the industry associations on the restructuring of the textile industry.[10] An equipment registration system for the nine branches of the textile industry was designed to restrain expansion and prevent new entries: "In order to prevent overheated equipment expansion competition from raising management uncertainty, this system, started in 1957 as a cartel based on the Smaller Industries Organization Law, has been exempted from the Anti-Trust Law."[11] Although abolished in the synthetic-fiber and natural-fiber spinning industries some time ago, MITI only recently bowed to U.S. pressure and announced abandonment of the system for smaller and medium-size spinners. Anxiety over excessive competition and hope for suitable profits suggest the dissatisfaction evident in Japan with unconstrained market dynamics and the high priority given in Japan to stability and predictability. Such concerns provide fertile ground for cooperative efforts to moderate market fluctuations.

Given the commitment to predictability and order, how does a capitalist society manage decline in unruly markets? Rogers J. Hollingsworth and Leon N. Lindberg describe states and firms as two forms of mediation that replace "market mechanisms as devices for the co-ordination and allocation of resources" and examine various blends of associative behavior among firms, and of government inter-

---

9. Matsumoto Koji, *The Rise of the Japanese Corporate System* (New York: Kegan Paul, 1991), 43.

10. Sen'i kōgyō shingikai (Committee on textile manufacturing) and Sangyō kōzō shingikai (Committee on industrial structure), "Kongo no sen'i sangyō no kōzō kaizen no arikata" (Directions for the coming structural improvement of the textile industry), *NBG*, December 1978, 9–21.

11. "Japan Abolishing Textile Equipment Registration by 1995," *JTN*, February 1992, 19–20.

vention to distinguish forms of economic governance.[12] William G.
Ouchi has written of three forms of mediation in capitalist societies—
markets, bureaucracies, and clans. According to Ouchi imperfections
of market dynamics can be moderated by the mediation of state
bureaucracies, or by the efforts among firms and industry associations
along the lines of Durkheim's organic solidarity. He concludes that
such kinship-like ties among firms in Japan helps shield them from the
vagaries of the market.[13]

Despite the well-publicized interventions by MITI in Japan's economic direction, I find that industry rather than government has
played the major role in the restructuring of the textile industry. Thus
I focus on the major textile firms and the business world of textile production and commerce. Labor groups are represented within firms,
and the state adopts and implements industrial restructuring programs only together with firms. Rodney Clark examines patterns of
firms within industries and finds a distinctive "society of industry" in
Japan. In a similar vein, Matsumoto describes Japanese society as "a
pluralistic welfare system having corporations as its basic units."[14]
Long-term contracting and subcontracting relations among firms, as
opposed to "spot contracting" based solely on price, help moderate
market dynamics. The same practice provides the best evidence of
Ouchi's clan-type relations among firms. Ron Dore discovers enduring ties between firms, which moderate the effects of disorderly markets, pointing to patterns of "relational contracting," which sustain
"networks of preferential, stable, obligated bilateral trading relationships."[15]

The story of textiles and transitions chronicles an effort to mediate
interests in an unruly market. Shutting down mills across Japan would

12. "The Governance of the American Economy: The Role of Markets, Clans, Hierarchies, and Associative Behavior," in *Private Interest Government: Beyond Market
and State*, ed. Wolfgang Streeck and Philippe C. Schmitter (London: Sage Publications,
1985), 221–55.

13. "Markets, Bureaucracies, and Clans," *Administrative Science Quarterly* 25
(March 1980): 129–40, and *The M-Form Society* (Reading, Mass.: Addison-Wesley,
1984).

14. Rodney Clark, *The Japanese Company* (New Haven: Yale University Press,
1979); Matsumoto, *Rise of the Japanese Corporate System*, 234.

15. "Goodwill and the Spirit of Market Capitalism," *British Journal of Sociology*
34, no. 4 (1983): 468, and *Structural Adjustment in Japan, 1970–1982* (London:
Athlone Press, 1986).

have brought widespread dislocation. Idle mills would have put workers on the streets, burdened the banks with defaults on credit lines, and wrought havoc on subcontractors and shareholders. Fears of shuttered mills mobilized industry leaders and public officials to work together to ease the transition in a declining industry, but reformers soon found themselves confronting private interests arrayed against the public interest in managed decline, as some firms within the industry bristled at the prospect of industry-wide patterns of adjustment. Disagreement soon surfaced between the majority of mogul firms representing the "industry" and a minority of dissenters and mavericks. Restructuring efforts only aggravated differences between state and firm, labor and management, and producer and supplier.

Mogul and maverick firms took different paths in a supposedly common adjustment process. Moguls complied with capacity reductions, mavericks balked. Widely recognized leaders in their textile industry, the moguls with their large workforces, extensive exposure with commercial banks, and close ties to Diet members were not likely to be sacrificed in any government-sponsored restructuring program. Established in lucrative markets and secure with a system of subcontractors and affiliates, the moguls were intent on maintaining their market share and profits at the very time they were investing in adjustment. The contrast with the smaller newcomers could hardly be more dramatic. Newcomers to the front line of textile production were less secure within the industry and smaller in scale; they cared less about maintaining the commonweal of an industry hierarchy to which they owed little allegiance, and more about expanding their own market niche.

Moguls and mavericks gambled for high stakes in the textile transition. The annual value of Japan's textile exports increased threefold in between 1970 and 1990, with the industry exporting textiles worth $2.4 billion in 1970, $6 billion in 1980, and $7 billion by 1990. But imports overshadowed exports. The annual value of imported textiles jumped from $1.2 billion in 1970 to $5.5 billion in 1980, and then *tripled* to $15.4 billion by 1990.[16] Competition with cheaper imports forced change at the textile firms, where the number

16. JCFA, *Sen'i handobuku 1992* (Textile handbook 1992) (Tokyo: Sen'i Sōgō Kenkyūkai, 1991), 70; JSA, *Statistics on the Japanese Spinning Industry 1991* (Osaka: Japan Spinners' Association, 1992), 100.

of operable spindles fell from a total of 16 million in 1960, to 9.6 million in 1990.[17] Total production of natural fibers in Japan declined by 50 percent between 1961 and 1990. Industry executives spoke publicly of a shift to multifiber production and to smaller lots of higher value-added products, but admitted privately that capital was fleeing textile production.[18] The immediate result was a reduction in the workforce.

Employment declined dramatically at the plants of both natural-fiber and synthetic-fiber producers. Take for example the recent history of three firms in the industry. Teijin Limited is a leading synthetics maker with a wide range of polymer technologies. Kanebo diversified, away from a concentration in synthetics and natural-fiber production to various consumer industries, particularly cosmetics and pharmaceuticals. Fuji Spinning remains mainly a manufacturer of natural and synthetic fibers.[19] Teijin Limited employed 15,700 in 1965, Kanebo 25,600, and Fuji 8,800.[20] Some mills were closed and workers let go. In 1992 employment at Teijin had fallen to 6,700, at Kanebo to 10,000, and at Fuji to 2,400.[21] Kanebo operated fourteen spinning mills in 1967, and only six in 1990. Fuji listed nine plants in 1967, and six in 1990.[22] And besides an industry-wide decline in production and employment, there was also a clear shift out of textile production at the firms. Synthetic fibers accounted for 93 percent of total sales at Teijin in 1971, but only 65 percent in 1992. Synthetics, cotton, and wool amounted to three-quarters of total sales at Kanebo in 1971, but only 32 percent in 1992. Synthetic and natural fibers amounted to 100 percent of total sales at Fuji in 1971, and fell only to 94 percent

17. JCFA, *Sen'i handobuku 1992*, 14. The JSA listed a total of only 6.5 million spindles in operation in 1990. *Statistics on the Japanese Spinning Industry 1991*, 31.

18. Production of synthetic fibers increased from 2.5 million tons to 3.3 million tons during the same period. JCFA, *Sen'i handobuku 1992*, 25.

19. "At Fuji Spinning Co. Ltd., which has placed the most emphasis on strengthening its production, about 70% of total yarn production is of fine count combed cotton yarn of 50s or above." "Japanese Spinners Feel Profitable in Their Fabric Exports Again," *JTN*, January 1989, 41.

20. Diamond Publishing, *Diamondo kaisha yoram 1965* (Diamond firm directory, 1965) (Tokyo: Diamond Publishing, 1965); Nihon Sen'i Kyōkai, ed., *Sen'i nenkan 1966* (Textile yearbook, 1966) (Tokyo: Nihon Sen'i Kyōkai, 1965).

21. Toyo Keizai, *Japan Company Handbook, First Section, Spring 1992* (Tokyo: Toyo Keizai, 1992), 236, 217, 221.

22. JSA, *Statistics on the Japanese Spinning Industry 1968* (Osaka: Japan Spinners' Association, 1969), and *Statistics on the Japanese Spinning Industry 1991*.

in 1992.[23] Fears of massive layoffs or of failure at the highly leveraged textile firms prompted extensive efforts to broker the adjustment among state, capital, and labor.

Results of the transition among leading firms indicate the relative merit of adjustment efforts. We find limited growth among the synthetics producers, survival among the cotton spinners, and some success among both groups in efforts to diversify into nontextile areas of sales and production. Annual sales at Teijin fell from Y450 billion in 1980 to Y325 billion ($2.3 billion) in 1990, but net profits rose from Y6 billion in 1980 to Y20 billion in 1990 ($144 million).[24] Sales at Kanebo have doubled from Y225 billion in 1980 to Y527 billion ($3.7 billion) in 1990. Kanebo rebounded from net losses in 1981 to net profits of Y3.4 billion ($24 million) a decade later. Sales at Fuji Spinning have remained steady across the decade, with Y85 billion in 1980 and Y86 billion ($600 million) in 1990. Improved labor productivity at the firm has led to expansion of net profits from Y186 million in 1980 to Y711 million ($5 million) in 1990.[25] Prosperity in some cases, survival in others, and the absence of numerous local newcomers to the market has made possible the continuing hegemony of the larger firms in the industry.

There are two contrasting themes evident in the adjustment effort. One is survival and persistence of hierarchy among filament, yarn, and fiber producers since World War II. The lineup of major textile firms has remained relatively stable in both market share and capital, apart from Toyobo's purchase of Kureha in 1966, and the merger of

23. Noyes Data Corporation, *Textile Industry of Japan, 1971* (Park Ridge, N.J.: Noyes Data Corporation, 1971); see also Dodwell Marketing Consultants, *Industrial Groupings in Japan 1978*, rev. ed. (Tokyo: Dodwell Marketing, 1978), 382, 381, and Toyo Keizai, *Japan Company Handbook*.

24. "Since the April 1982 term when the equity method was revised, an increasing number of companies have been adopting the equity method and issuing a consolidated report in order to reflect the earnings of affiliated companies in which their equity is 20%–50%." Toyo Keizai, *Japan Company Handbook*, 29. Teijin reported on both a consolidated and nonconsolidated basis. On a consolidated basis, net sales amounted to Y612 billion ($4.3 billion), and operating income to Y45 billion ($321 million). I refer only to nonconsolidated financial reports, since the latter have appeared more consistently across the years of change in the industry.

The year "1990" refers to the fiscal year ending 31 March 1991. Throughout this book I calculate all dollar amounts of yen figures by the same 31 March 1991 exchange rate of U.S. $1 = Y141.

25. Diamond Publishing, *Diamond's Japan Business Directory 1988* (Tokyo: Diamond Publishing, 1988); Toyo Keizai, *Japan Company Handbook*.

Table 1. Spinning moguls, 1990
(assets and sales expressed in billions of U.S. dollars)

Firm	Employees	Assets	Sales
Toyobo	8,001 (−.39)[a]	4.0 (+.86)	2.4 (+.28)
Kanebo	10,376 (+.93)	4.7 (+1.1)	3.7 (+1.0)
Unitika	5,479 (−.10)	3.3 (+.74)	2.0 (+.36)
Fuji Sp.	2,380 (−.41)	0.5 (+.24)	0.6 (+.01)
Nisshinbo Industries	6,166 (+.07)	2.6 (+2.7)	1.4 (+.25)
Kurabo Industries	3,501 (+.01)	1.1 (+1.3)	0.9 (−.07)
Daiwabo	2,883 (−.56)	1.1 (+.44)	0.5 (−.22)
Shikibo	2,706 (−.21)	0.9 (+.50)	0.4 (.00)
Nitto Boseki	4,321 (−.22)	1.4 (+1.3)	1.0 (+.17)
Omi[b] Kenshi	3,341 (−.24)	0.2 (−.21)	0.4 (+.01)

Sources: Employment and total assets reported for March 1991 and March 1981. Net sales listed for fiscal years 1980 (i.e., 1 April 1980–31 March 1981) and 1990. Data for fiscal year 1 April 1980–31 March 1981 are drawn from *Kaisha shikihō, 57 nen, shinshun* (Company reports, Spring edition, 1982) and from *Nikkei eigyō shihyō 1982* (Nikkei financial analysis, Spring 1982). Data for fiscal 1990 are from unconsolidated financial statements cited in the *Japan Company Handbook, First Section, Spring 1992.*

[a]Ratios in parenthesis represent comparison with a base of fiscal year 1980. Employment at Toyobo, for instance, declined 39 percent from the base-year total of 13,236 workers in 1980 to only 8,001 workers in 1990. Assets in the same decade increased 86 percent from a level of about $2.16 billion in 1980 to $4 billion in 1990. Note that Toyobo, Kanebo, and Unitika are listed as both major synthetic-fiber makers and major spinners.

[b]Data for Omi Kenshi are drawn from *Sen'i fashiun nenkan 1992,* 329, and from Nihon Keizai Shimbunsha, *Nihon eigyō shihyō 1981* and *Nihon eigyō shihyō 1990.*

Nichibo and Nippon Rayon to form Unitika in 1969. The top sixteen textile firms include the spinning firms listed in Table 1, and synthetic-fiber makers listed in Table 2. The three largest spinners imported polymer technology in the late 1950s and 1960s. Toyobo, Kanebo, and Unitika (i.e., Nippon Rayon) subsequently emerged as both natural-fiber spinners and synthetic-fiber makers. The other seven firms among the large spinners remained dedicated cotton spinners for the most part, though all have diversified into polyester/cotton blends, and also into nontextile products. These seven spinners are Nisshinbo

Table 2. Mogul synthetic-fiber makers, 1990
(assets and sales expressed in billions of U.S. dollars)

Firm	Employees	Assets	Sales
Teijin	6,777 (−.06)[a]	6.6 (+1.1)	2.3 (−.27)
Toray	10,428 (−.24)	8.1 (+1.2)	4.1 (+.10)
Toho Rayon	1,748 (−.24)	0.6 (+1.3)	0.5 (+.08)
Mitsubishi Rayon	5,145 (+.45)	3.4 (+1.8)	1.9 (+.46)
Kuraray	6,123 (−.01)	2.9 (+.93)	1.9 (+.35)
Asahi Chemical Industry	15,547 (+.14)	9.1 (+1.0)	6.8 (+.62)
Toyobo	8,001 (−.39)	4.0 (+.86)	2.4 (+.28)
Kanebo	10,376 (+.93)	4.7 (+1.1)	3.7 (+1.0)
Unitika	5,479 (−.10)	3.3 (+.80)	2.0 (+.36)

Sources: Please see the source note to table 1.

[a]Ratios in parenthesis represent comparison with a base of fiscal year 1980. Employment at Teijin, for instance, declined 6 percent from the base-year total of 7,227 workers in fiscal 1980, to only 6,777 workers in fiscal 1990. Assets in the same period increased 110 percent from a level of about $3.1 billion in 1980, to $6.6 billion in fiscal 1990. Note that Toyobo, Kanebo, and Unitika are listed as both major synthetic-fiber makers and major spinners.

Industries, Kurabo Industries, and Fuji Spinning, Daiwabo, Shikibo, Nitto Boseki, and Omi Kenshi. As for the chemical and synthetic filament and fabric sector, the big synthetic-fiber producers are Asahi Chemical Industry, Toray Industries, Teijin, Mitsubishi Rayon, Kuraray, Toho Rayon, Toyobo, Kanebo, and Unitika. I refer to these synthetic-fiber producers, as well as the spinners above, as "moguls" to denote their size and high status among the larger group of twenty-two polyester producers and sixty midsize and larger spinners. Moguls dominate the industry both in the market and at the leading industry associations.

A small group of resisters has emerged to oppose the moguls. Even a few mogul firms found it difficult to support the suggested restructuring of the industry. Omi Kenshi, Nisshinbo, and Nisshinbo's affiliate Toho Rayon broke ranks with their fellow moguls. The "dissenters" balked at the industry policy of reducing spindle capacity from the late 1960s, and instead maintained spinning capacity, modernized their equipment, and even added some spindles.[26] Upgrading of existing capacity can be as important as expansion for increasing value-added production. Nisshinbo has far more spindles today with doubling capacity than it did in 1968, and Omi Kenshi has both spindles with doubling capacity and a few more recent open-ended spindles. Toho Rayon has added both open-ended spinning capacity and doubling capacity.[27] The commitment of the dissenters to cotton spinning has jostled but not upset the hierarchy of producers in cotton production. In the decade between the end of 1971 and the beginning of 1982, Nisshinbo moved from the fifth to the fourth rank in capacity among the twelve spinners. By 1991, Nisshinbo had climbed to third place, trailing only Tsuzuki and Toyobo. Omi Kenshi jumped from eleventh to eighth in the rankings between 1971 and 1982, and was up to fourth place by 1991. Toho Rayon moved up from the bottom of the ladder of top producers in 1971, to tenth place by 1991.[28] A choice for limited expansion and upgrading among the three dissenting moguls appears in hindsight more stubborn than defiant. They simply chose not to reduce natural-fiber production, investing instead in upgrading and limited expansion to maintain their share of the natural-fiber market.

Unlike the dissenters, the newcomers listed in Table 3 such as Tsuzuki Spinning and Kondo Spinning of Nagoya seized the opportunity provided by industry-wide reductions in capacity to expand aggressively. Tsuzuki has been adding capacity since the early seventies. Kondo expanded within Japan between 1970 and 1980 and then decreased the number of spindles at local mills over the last decade. Tsuzuki reported about 400,000 spindles in production in 1968 and

26. Nisshinbo owns 24.4 percent of the total shares in Toho Rayon and retains four positions on its board. The chair of the board, the president, a managing director, and an auditor are all Nisshinbo employees posted to Toho Rayon. Okurashō (Finance Ministry), *Yūka shōken hōkokusho sōran—Toho Rayon Kabushiki Kaisha, 1991* (A compendium of financial reports—Toho Rayon Company, Ltd., 1991) (Tokyo: Okurashō, 1991). Cotton yarn represented about one-fifth of the total annual value of production at Toho Rayon in 1990, and better than a quarter of total annual sales.

27. Toho Rayon also increased both the scale and quality of its cotton spinning. The firm reported 260,000 spindles in 1968 and 308,000 in 1991. The latter figure included 9,300 open-ended frames and 12,000 spindles with doubling capacity. JSA, *Statistics on the Japanese Spinning Industry 1968* and *Statistics on the Japanese Spinning Industry 1991*.

28. Yoshioka Masayuki, *Sen'i* (Textiles) (Tokyo: Nihon Keizai Shimbunsha, 1986), 73; JSA, *Statistics on the Japanese Spinning Industry 1991*.

Table 3. Maverick spinners, 1990
(sales expressed in billions of U.S. dollars)

Firm	Employees	Sales
Tsuzuki Spinning	4,683 (+.25)[a]	0.907 (+.65)
Kondo Spinning	2,100 (−.34)	0.409

Sources: Data on employment and net sales reported in the *Sen'i nenkan* (Textile yearbook) for 1982 and 1992, reporting data on fiscal years 1980 and 1990. The two firms reported net sales only in the 1992 volume. I obtained figures for net sales in fiscal 1980 from a publication provided by Tsuzuki Spinning titled *Aggressively Tsuzuki.* I could not obtain data on net sales for Kondo Spinning in fiscal 1980.

[a]Ratios in parenthesis represent comparison with a base of fiscal year 1980. Employment at Tsuzuki, for instance, increased about 25 percent from the base-year total of 3,500 workers in fiscal 1980, to 4,683 workers in fiscal 1990. Sales increased 65 percent from a level of about $546 million in fiscal 1980 to $906 million in fiscal 1990. The privately held Tsuzuki and Kondo firms do not publish data on total assets.

800,000 in production by 1991. The latter included 18,000 open-ended frames and 10,000 with doubling capacity. Kondo reported 280,000 spindles in 1968, 450,000 in 1980, and 370,000 in 1991, including 5,000 open-ended frames.[29] Tsuzuki increased its workforce from 3,300 employees in 1966 to 5,130 in 1990, with Kondo reporting 3,700 employees in 1966, but only 2,100 in 1990. Tsuzuki moved from ninth to third among cotton producers between 1971 and early 1982, and to first place by 1990, operating 200,000 more spindles than its nearest competitor. Kondo Spinning ranks fifth in capacity among the largest cotton spinners today.[30]

We can assume the two maverick firms produce about a third of the cotton yarn and fiber manufactured in Japan, since Tsuzuki and Kondo annually bring in about a third of all raw cotton imported.[31] Tsuzuki reported sales of Y77 billion in 1980 and Y130 billion ($928 million)

29. JSA, *Statistics on the Japanese Spinning Industry 1968* and *Statistics on the Japanese Spinning Industry 1991.* The term "newcomer" is a translation of the Japanese term *shinbō,* or "new spinner." Although Tsuzuki was founded as a weaver in the early 1940s, its postwar reorganization from weaving to spinning, and relatively late membership in the JSA, has left it with the status of newcomer. In contrast, the moguls with their prewar history as major spinning firms, and early postwar registration as members of the JSA, are known as the "major spinners," or *ōte kaisha.*

30. Yoshioka, *Sen'i,* 75; JSA, *Statistics on the Japanese Spinning Industry 1991.*

31. The scale of their purchase of raw cotton was confirmed by officials of the Japan Cotton Traders' Association, moguls, traders, and officials of the Raw Material Department at Tsuzuki headquarters in Nagoya.

in 1989. Kondo reported annual sales of Y250 million in 1980 and Y57.2 billion ($406 million) in 1989.[32] I refer to Tsuzuki Spinning and Kondo Spinning as "mavericks" because of their defiance of mogul direction in the restructuring process, but not as "free riders," for both firms remain active in the JSA. Tsuzuki also belongs to the powerful Japan Chemical Fibers' Association (JCFA). Government officials, mogul firms, and industry association officials with access to production and equipment statistics from the two member firms were fully aware of the divergence of the two Nagoya spinners. Maverick behavior raises important questions about interest mediation in the adjustment process. Why did the mavericks and dissenters deviate from the state/industry consensus on capacity reductions? Why did this deviation not destroy the common direction or compass of change? Why did this dissent not dissolve the consensus among moguls?

Transitions

Most of the firms in Japan's textile industry reduced production and restricted themselves to "appropriate competition as part of an effort stretching three decades from 1956 to restructure an industry in decline. Some did not. I suggest the liberal corporatist strategies invoked to effect "transition" sustained the hierarchy of moguls without discouraging the defection of a few newcomers. The Japanese government and leaders in Japan's textile industry planned an industry adjustment that would benefit society, the industry, and the individual firm. Advisory councils (*shingikai*) organized under the aegis of MITI formulated and refined "visions" for the industry. The *shingikai* encouraged vertical integration within the industry, product specialization, diversification, and offshore investment to sustain a local industry despite its declining share of domestic textile sales. The government was to help with subsidies or loans for scrapping machinery in the industry and with exemptions from antitrust laws forbidding production cartels. Most of the mogul firms complied with such initiatives,

---

32. Nihon Sen'i Kyōkai, ed., *Sen'i nenkan* (Textile yearbook), 1980, 1990. Annual sales from 1965 are also listed on page 16 of *Aggressively Tsuzuki*, a publication of Tsuzuki Spinning.

reducing their spinning capacity by at least 50 percent since the late 1960s, and specializing, diversifying, and moving production off-shore. What draws our attention, however, are the exceptions—dedicated spinners who maintained or increased capacity.

The question of consensus and dissent draws us beyond a simple examination of the adjustment procedures to an investigation of the formation of adjustment priorities. John R. Bowman has written of the need for collective action among firms to ensure the necessary conditions for productive competition.[33] His emphasis on the necessity of cooperation evident in the history of the U.S. coal industry offers an excellent starting point for studies of industrial restructuring. Bowman discusses "free riders" who defected from price-control efforts. The Japanese firms were not "free riders," but rather "mavericks" who were generally good citizens within the industry and carefully paid the costs of divergence from industry norms of adjustment. Mavericks would, for example, purchase the registration papers for scrapped equipment from other firms. They could then expand by adding equipment on the basis of these receipts for idled equipment. Mavericks also paid higher dues at the main industry associations, since the fee structure is based on cotton spinning capacity and on annual production. As they rose in both categories, they took on proportionally larger shares of the costs of the industry's cooperative efforts. So, "free riders" they were not, but neither did they meekly follow the pattern of cooperation in capacity reductions that characterized adjustment efforts among the larger firms in the textile industry.

The transition in the Japanese textile industry highlights patterns of cooperation. Firms shared information on technology, raw materials, and markets and jointly negotiated agreements on import growth with textile producers in competing nations and on state support for common directions of change. The umbrella offered some shelter from the vagaries of market fluctuations and permitted some stability in market share. Scholars have long emphasized such patterns in Japan's polity and economy. Ishida Takeshi has written of conformity and competition in Japan's recent political and economic history: "Competition in loyalty makes it possible to reconcile the principle of competition's inherent conflict with conformity, and to make competition a vital feature of the conformity-oriented national

33. *Capitalist Collective Action* (Cambridge: Cambridge University Press, 1989).

group."[34] Murakami Yasusuke has argued that solidarity based on the *ie* or household is the key to cooperation among state, industry, and firm.[35] A recent study of cooperation and competition among state and industry in information technologies suggests the state effectively reduced uncertainty in efforts to develop technology by providing a cooperative umbrella of joint research funding with the industry.[36]

Our task is to uncover the underlying structures that make such cooperation possible. Margaret A. McKean has singled out encompassing interests as the key to successful restructuring in Japan. Unlike in Europe and the United States, groups in Japan find ways to extend their special interests to align with the broader interests of society, and with the long-range time frames necessary for societal change. She cited an expansion of corporatist structures in Japan recently, stretching special interests into more encompassing, collective interests.[37] The textile industry is a good example of this phenomenon. But despite collective action under a government umbrella, some firms did not conform to the "consensus" vision. One might well expect such dissent to destroy common patterns of change, as it did in other declining industries in Japan.[38] Why in the textile industry did it *not*?

The corporatist thesis provides us a hypothesis for testing patterns of collective action. Under corporatism individual actors realign their particular interests to ensure the long-range advantages of designated, organized interest groups. Unruly markets, for example, are moderated (under corporatism) by firms within industries, as well as by cooperation between management and labor and between firms and

34. *The Integration of Conformity and Competition—A Key to Understanding Japanese Society* (Tokyo: Foreign Press Center, 1980), 11. See also his related article, "Conflict and Its Accommodation: Omote-Ura and Uchi-Soto Relations," in *Conflict in Japan*, ed. Ellis S. Krauss, Thomas P. Rohlen, and Patricia G. Steinhoff (Honolulu: University of Hawaii Press, 1984), 16–38.

35. "Ie Society as a Pattern of Civilization," *Journal of Japanese Studies* (Summer 1984): 360.

36. Martin Fransman, *The Market and Beyond* (Cambridge: Cambridge University Press, 1990).

37. Margaret A. McKean, "State Strength and the Public Interest," in *Political Dynamics in Contemporary Japan*, ed. Gary D. Allinson and Sone Yasunori (Ithaca: Cornell University Press, 1993), 72–104.

38. Gregory Noble discussed cases of divisive dissent in "The Japanese Industrial Policy Debate," in *Pacific Dynamics: The International Politics of Industrial Change*, ed. Stephan Haggard and Chung-in Moon (Boulder, Colo.: Westview Press and CIS-Inha University Press, 1989), 53–95.

state bureaucracy. Such long-established and carefully structured patterns of negotiations between capital and labor, and between capital and the state, are evident in the "substitution of bargaining for market or politics" in this sector of Japanese industry.[39] In the textile industry interests are mediated through institutionalized, continuous consultation between interest organizations, state bureaucracies, and political parties.[40] That actual bargaining takes place at least between capital and the state distinguishes the "neo" or "liberal" or "societal" corporatism in Japan from European fascism and from recent examples of authoritarian or "state" corporatism.[41]

Noel Sullivan has recently criticized corporatism theorists for diverting attention from issues of democratic processes. He insists that focusing on intermediation or "concertation" overlooks inequalities between stronger and weaker interest groups. This is a caveat to keep in mind as we examine the shifts in balance of power between state and capital, moguls and mavericks, and capital and labor in the restructuring of the textile industry in Japan. Sullivan's criticism echoes Charles Anderson's assertion that "structured relationships between important interests and the state take on a life of their own, become dominant over the formal institutions of representative democracy, and are extremely resistant to change."[42] Anderson's con-

39. Ron Dore, "Japan: A Nation Made for Corporatism?" in *Corporatism and Accountability: Organized Interests in British Public Life,* ed. Colin Crouch and Ronald Dore (Oxford: Clarendon Press, 1990), 61, and "How Fragile a Super State," in *Japan and World Depression, Then and Now,* ed. Ron Dore and Radha Sinha (London: Macmillan, 1987), 98–103.

40. On corporatist negotiation of interests, see Reginald L. Harrison, *Pluralism and Corporatism* (London: George Allen & Unwin, 1980), 64. Also see Peter Katzenstein, *Small States in World Markets: Industrial Policy in Europe* (Ithaca: Cornell University Press, 1985), 24.

41. Philippe Schmitter explains that "societal" corporatism differs from "state" corporatism because it comes "into existence largely, but not exclusively, as the result of inter-associational demands and intra-organizational processes—from below, so to speak, rather than from a conscious effort by those in power to mold the type of interest intermediation system most congenial with their authoritarian mode of domination." "Interest Intermediation and Regime Governability in Contemporary Western Europe and North America," in *Organizing Interests in Western Europe: Pluralism, Corporatism, and the Transformation of Politics,* ed. Suzanne Berger (Cambridge: Cambridge University Press, 1981), 292.

42. Noel O'Sullivan, "The Political Theory of Neo-Corporatism," in *The Corporate State: Corporatism and the State Tradition in Western Europe,* ed. Andrew Cox and Noel O'Sullivan (London: Edward Elgar, 1988), 12–13; Charles W. Anderson, "Political Design and the Representation of Interests," *Comparative Political Studies* 10, no. 1 (1977): 144.

cern alerts us to the deeper, "institutionalized" character of such relationships, and indeed, organized interests in the textile industry in
Japan appear both extended and embedded. The tendency in the textile industry to extend special interests into more comprehensive interests is embedded rather than imposed, which facilitates the
development of societal corporatist styles of negotiation.

I argue that corporatist strategies of adjustment permit dissent without destroying consensus, but admit that the concept itself of "Japanese corporatism" remains problematic. If I cite corporatism as an
answer to the riddle of dissent and consensus, the corporatist hypothesis itself becomes the question. Is it corporatism, or perhaps patterned pluralism, or simply reciprocity that distinguishes the
adjustment in textiles? Some would argue rather for a pluralist style of
interest negotiation, or perhaps a more distinctly Japanese style of
mutual reciprocity between leading interest groups. In a recent study
of ties between business and the state in the energy sector, Richard
Samuels discovered patterns of "reciprocal consent" between firm and
government, and wrote of "a mutual accommodation of state and
market."[43] His work has helped turn my attention to an emerging balance of power in the textile industry between state and firm that reinforces established patterns of negotiation and confirms existing
hierarchies in order to maintain stability and growth.

Various observers have remarked on the distinctive patterning of
political ties within contemporary Japan, especially in relations
between government and the business community. The consensus
appears to fall between Western-style pluralism and more corporatist
forms of interest mediation. For instance, Muramatsu Michio and
Ellis S. Krauss have recently proposed the thesis of "patterned pluralism" to distinguish the Japanese process of political bargaining
between the Diet and the wider society. Despite acknowledging the
existence of "consistent coalitions of actors with relatively predictable
degrees of influence on policymaking," they find that the system pos-
sesses sufficient flexibility to merit the term "pluralistic."[44] Aoki

43. Richard J. Samuels, *The Business of the Japanese State: Energy Markets in
Comparative and Historical Perspective* (Ithaca: Cornell University Press, 1987).

44. Muramatsu Michio and Ellis S. Krauss, "The Conservative Policy Line and the
Development of Patterned Pluralism," in *The Political Economy of Japan*, ed. Yamamura Kozo and Yasuba Yasukichi, vol. 1, *The Domestic Transformation* (Stanford:
Stanford University Press, 1987), 538. See also Gary D. Allinson, "Politics in Contemporary Japan: Pluralist Scholarship in the Conservative Era—A Review Article," *Journal of Asian Studies* 48 (May 1989): 324–32.

Masahiko approaches the same problem with a more limited focus, looking to issues of interest representation and bureaucratic regulation in the ministries of the Japanese government and offering the hypothesis of "administered pluralism."[45] The term "administered pluralism" connotes two contrasting functions for the bureaucracy: representation of specific interest groups on a long-term, continuing basis and more autonomous regulation of the industry according to the norms of economic rationality and political feasibility. The need to distinguish between patterned pluralism and administered pluralism reflects the difficulty of bringing the Japanese political economy into a comparative context along a continuum of pluralist interest competition and corporatist interest intermediation. Both theses provide insight into the complexity of interest bargaining in Japan.

I find more evidence for societal corporatism than for patterned pluralism in the actions of the Japanese textile industry. The interdependence of the state and private-interest organizations evident in structured, long-term negotiating relationships distinguishes corporatism from pluralism. Gerhard Lehmbruch lists three traits of corporatism: (1) a network of centralized, structured interest groups; (2) privileged, institutionalized linkages between such groups and government; and (3) a tripartite "social partnership" among capital, labor, and government.[46] Business associations and labor federations in Japan offer evidence of the first two traits; and coordination of policy and practice across labor, capital, and government in the textile transition, evidence of the third. Alan Cawson recently offered a concise definition of corporatism: "Corporatism is a specific socio-political process in which a limited number of monopolistic organizations representing functional interests engage in bargaining with state agencies over public policy outputs."[47] In exchange for favorable government policies, the leaders of the interest organizations agree to ensure the cooperation of their members. A focus on interests that are "struc-

45. "The Japanese Bureaucracy in Economic Administration: A Rational Regulator or Pluralist Agent," in *Government Policy towards Industry in the United States and Japan*, ed. John B. Shoven (Cambridge: Cambridge University Press, 1988), 267–68.

46. "Concertation and the Structure of Corporatist Networks," in *Order and Conflict in Contemporary Capitalism*, ed. John H. Goldthorpe (Oxford: Clarendon Press, 1984), 61.

47. "Corporatism," in *The Blackwell Encyclopedia of Political Thought*, ed. David Miller (New York: Basil Blackwell, 1987), 105.

turally privileged" in the shaping of state policy draws attention to issues of group power and their implications for both democracy and capitalism.[48]

One student of Japanese corporatism concluded that corporatist strategies should be analyzed with a focus solely on direct relations between the state and individual firms because business interest associations have proved less and less capable in mediating industry interests to the government.[49] I find in the textile industry, however, that the associations continue to play a major part in mediating relations between state and industry. I also find industry associations prominent in representing the interests of an "industry" to the individual firms. In this book I examine two monopolistic industry associations within the Japanese textile industry, the Japan Spinners' Association (JSA) and the Japan Chemical Fibers' Association (JCFA), and the parallel organization that labor has developed to press its interests, the Zensen Federation.

T. J. Pempel and Tsunekawa Keiichi have drawn attention to the exclusion of Japanese labor in corporatist mediations.[50] I found that labor's interests were taken into account in the restructuring plans for the textile industry, despite labor's subordinate status in negotiations with capital and the state. Two factors, however, weaken labor's bargaining position: the identification of workers with their enterprise rather than with a craft union and the alliance between labor and minority parties. Yet labor has found a role in the restructuring of the Japanese textile industry to this point, even without an autonomous voice in shaping the industry-wide vision directing the transition. I will look at what labor's subordinate position implies both for labor itself and for ties between state and capital. Pempel and Tsunekawa also emphasize the erosion of corporatist ties due to the fading depen-

48. Peter J. Williamson, *Varieties of Corporatism* (Cambridge: Cambridge University Press, 1986), 163.

49. George Aurelia, *The Comparative Study of Interest Groups in Japan: An Institutional Framework* (Canberra: Australian National University, Australia-Japan Research Centre, 1982), 66.

50. "Corporatism without Labor? The Japanese Anomaly," in *Trends toward Corporatist Intermediation*, ed. Philippe C. Schmitter and Gerhard Lehmbruch (Beverly Hills, Calif.: Sage, 1979), 231–70. See also Tsujinaka Yutaka, "Gendai Nihon seiji no kooporatizumu-ka" (The shift to corporatism in Japanese politics), in *Kōza seijikagu* (Introduction to politics), ed. Uchida Man, vol. 3, *Seiji katei* (The political process) (Tokyo: Sanrei Shobo, 1986), 223–62.

dence of industries on government support and the expansion of international ties. The point is well taken and reminds us that corporatism is indeed an ideal type in desperate need of context and specification.

An ideal type of corporatism suggests variation along a continuum of corporatism, not only across societies, but also across industries within a society, and across time within an industry.[51] T. J. Pempel observes that "though Japan is distinctly pluralistic, there has been a consistently close relationship between the state, bureaucracy, ruling politicians, and big business."[52] The textile industry is no exception. One purpose of this book is to specify neocorporatist patterns of policy formation and enforcement evident in the restructuring of the industry since 1956. I would emphasize at the outset that such neocorporatist strategies are not only structured and institutionalized, but also hierarchical. I would agree also with Murakami Yasusuke that such corporatism in Japanese society tends to be inclusionary, including labor and capital in the textile industry, though hardly with comparable leverage in policymaking.[53] The prominent role of neocorporatist strategies of adjustment balancing dissent within a common compass of change is the major finding of this book.

Thesis

Two prominent themes in the restructuring of the Japanese textile industry are "adjustment" and "interest mediation." "Adjustment" refers to changes within the industry, that is, to the survival or failure of individual firms, the stability of the hierarchy, and to the emergence of newcomers. "Interest mediation" refers to the actions of the indus-

---

51. Aurelia emphasizes that "levels of corporatisation are, therefore, correlated with levels of direct government intervention and assistance in the economy." *Comparative Study*, 63, 66. Wilensky suggested variation among issues in Japan as well. He found more corporatist styles of negotiation in economic and industrial policy which drew closer attention from leading power groups in Japan, than in social and environmental policies. See Harold L. Wilensky, *Democratic Corporatism and Policy Linkages* (Berkeley, Calif.: University of California Press, Center for East Asian Studies, 1987), 13.

52. "Japanese Foreign Economic Policy: The Domestic Bases for International Behavior," *International Organization* 31 (1977): 733.

53. See Murakami Yasusuke, "The Japanese Model of Political Economy," in *The Political Economy of Japan*, ed. Yamamura Kozo and Yasuba Yasukichi, vol. 1, *The Domestic Transformation* (Stanford: Stanford University Press, 1987), 70.

try as a whole, to include consensus between the state and industry moguls on strategies of specialization, diversification, and offshore investment. "Mediation" can also refer to divergence from state-supported capacity-reduction programs and from production cartels. I trace the progress of the restructuring of the Japanese textile industry and assess the implications of dissent within consensus for theories of corporatism and collective action in Japan. Direction and divergence provide the focus for two chapters introducing the textile industry and selected moguls and mavericks. Consensus and dissent in turn draw our attention in subsequent chapters on state, association, and labor. Two later chapters on merchants rather than makers focus on mediation of both discrepant interests and divergent functions in the textile transition. Among the merchants, purveyors of fashion offer a foil to corporatist patterns of change among the upstream producers. Fashion houses offer little evidence of corporatist cooperation, but provide reliable market channels critical for corporatist styles of adjustment among their upstream suppliers.

The thesis of corporatist adjustment points to deeply rooted, enduring patterns of interest mediation supporting cooperative strategies of change. Opponents of the corporatist hypothesis suggest alternative explanations such as "dependence" or "structural dispersion" to account for the cooperation that has occurred during this period of transition. Executives at some textile firms favor the dependence thesis. They argue that cooperation has been forced on the industry because as an industry it is vulnerable to market dynamics beyond Japan's borders. This is true for raw materials. Japan's textile industry is totally dependent on imports of raw cotton, and partially dependent on imports of the chemicals necessary for chemical (i.e., rayon and acetate) and synthetic-fiber (i.e., acrylics, nylon, and polyester) production. In addition, the latter firms draw supplies from the local petrochemical industry, which depends on imports of foreign oil. A sharp rise in the price of oil immediately affects the cost of production of chemical and synthetic fiber. Such dependence on raw materials from abroad leaves the industry exposed to fluctuations in both supply and demand in foreign markets.

A similar dependence on basic materials can be found in spinning, but with less concentration among the firms. Fluctuation in the volatile raw-cotton futures market in New York does affect production costs for the Japanese cotton mills immediately. The midstream sector, including spinning, weaving, knitting, and twisting or dyeing, does tend to be concentrated in the ten mogul spinning firms, but we also find a few mavericks, plus a group of some thirty or so midsize producers, and still other, smaller mills engaged in this phase of production. Mogul spinners have made some headway in efforts to integrate this midstream production within their own firms, but still must rely on a network of subcontractors for weaving and knitting much of their yarn. We find that the industry becomes even more diffuse as we move "downstream" to the wholesaling and distribution of finished apparel. Larger textile mills must turn to a network of smaller subcontractors and wholesalers to actually make up garments and market their products. But even with only limited integration of the production process within their own firms, the moguls retain the greatest leverage in industrial policy formation among the various firms along the production line. Scale and function keep the major mills at the top of the pyramid in textiles, for the industry depends ultimately on large-scale producers of filament, fiber, and yarn.

The dependence argument in itself does not explain cooperation within the Japanese textile industry, though dependence itself may well encourage more communal strategies. The structural dispersion argument suggests that some more powerful authority is coordinating or even imposing cooperation on an unruly array of midsize and larger spinning mills. But we find no evidence of unilateral imposition of policy (state or otherwise) on a recalcitrant industry. Dispersion of assets and facilities among a range of larger and midsize firms in spinning, however, does make corporatist styles of cooperation extremely complex. Students of structural adjustment have pointed out the advantages of oligopolies, such as fewer players with a stronger common interest in adjustment.[54] Such an exclusive oligopoly does not exist in the textile industry; hence common strategies of restructuring are far more difficult to coordinate within industry associations.[55] Unlike the steel or aluminum industries, the textile industry in Japan is far more dispersed, despite the scale and leadership of the mogul firms. Whereas we might expect more cooperative, "corporatist" pat-

54. Susan Strange and Roger Tooze, *The International Politics of Surplus Capacity* (London: George Allen & Unwin, 1981), 82–83.

55. See Mark Campbell Tilton, "Trade Associations in Japan's Declining Industries: Informal Policy-Making and State Strategic Goals" (Ph.D. diss., University of California at Berkeley, 1990).

terns of negotiation between capital and state in oligopolistic sectors such as steel, similar strategies would be far more difficult to coordinate in a more dispersed industry such as textiles. Corporatism would suggest a monopoly role for employers' organizations such as the JSA or the JCFA, a fusion of their roles of representation and policy implementation, and a prominent role for the state in licensing monopoly representation and in codetermining policy. An argument for corporatism in the textile industry must give evidence of a prominent role of industry associations in the structured intermediation of interests between capital and the Japanese state, despite limited dispersion of productive capacity among larger and smaller mills in the spinning sector of the industry.

The story unfolds in three directions. I trace changes at leading firms, including efforts at integrated production, and especially at product specialization, offshore investment, and diversification. I explain survival at the mills, persistence of hierarchy among the firms, yet the rise of dissenters and mavericks. I sift through underlying networks and patterns of competition and cooperation evident in disputes and accommodations between moguls and mavericks, labor and management, mills and merchants across three decades. The Industrial Restructuring Council for Textiles crafted the "vision" statements that offered a common framework for change and provide the setting in which industry leaders, union officials, MITI planners, traders, and apparel makers are the central actors. I have culled a narrative of change from interviews with the actors. I bring to their story an analytic framework regarding interests, an argument regarding mediation, and a thesis about adjustment. I suggest that neocorporatist strategies prominent in the restructuring of the textile industry moderated decline, preserved hierarchy, and yet did not preclude emergence of newcomer firms.

2 The Textile Industry

Textile mills in Europe, the United States, and Japan have all faced declining market share since the early seventies. U.S. firms improved efficiency through horizontal mergers and garnered state support for adjustment with trade protections, whereas European firms looked less to protection from imports and more to capacity reduction, with state relief for individual firms. A third path was taken in Japan, where industry and state joined forces to direct change with sector-specific, long-term efforts to scrap excess capacity and modernize production.[1] A melding of special, specific interests with a wider public interest continues across the entire adjustment process. Planned adjustment in Japan suggests efforts to manage change as a group, but it does not preclude response by individual firms to market dynamics rather than to policy dictates.

In Japan efforts to shape encompassing interests from the special interests of capital and labor were evident in state policy. But for "capital" to achieve their interests, the interests of various firms had to be made into the common interest of an industry association. For

---

1. Careful planning appears one of the features of Japan's industrial restructuring process: "Japan's reduction in production reflected the country's conscious decision to cut back its textile industry." ILO, *General Report of the Twelfth Session of the Textiles Committee, 1991: Report I* (Geneva: ILO, 1991), 69. We read elsewhere that "the most consistent example of sector-specific policy has been in Japan, where the main emphasis has been on a combination of scrapping and modernization." OECD, *Structural Problems and Policies in OECD Countries: Textile and Clothing Industries* (Paris: OECD, 1983), 114. Britain in time turned to a sector-specific approach, as did other European nations. See also Brian Toyne et al., "The Textile Complex and the Textile Mill Products Industry," in *The Global Textile Industry* (London: Allen & Unwin, 1984), 116, 158–59, and Robin Ansan and Paul Simpson, *World Textile Trade and Production Trends*, Special Report #1108 (June 1988) (London: The Economist Intelligence Unit, 1988), 136–42.

"labor" to achieve their interests, the interests of various enterprise unions had to be made into a common labor interest. Closer to market, the challenges were more practical; there, mills had to shape common interests with merchants.

The story of change in textile manufacture begins in the late nineteenth century. Here we discover early patterns of cooperation and competition, of common direction yet divergence among firms within an industry, and of early instances of corporatist bargaining between state and capital. A chronicle of the industry introduces us to firms and to their associations which helped mediate the interests of capital with both state and labor. The effectiveness of such mediations depends in part on the structure of the industry. Industries of only loosely related firms or of firms of many different sizes and interests did not submit easily to compromise or effective mediation, whereas oligopolies found common ground quickly. The effectiveness of such mediations also depended on firms within an industry. Limited structural concentration supported cooperative efforts in the upstream sectors of the textile industry, but more important was a tradition of sorting out the interests that shaped the identities of particular firms within it.

Interest competition and cooperation between labor and capital and between state and capital took place in individual firms. Firms within industries (gyōkai) provide a stable forum in which interests among capital, labor, and the state can be negotiated. Murakami outlines a network of firms within industries distinguishing the Japanese economy and writes of the contributions of state industrial policy to the emergence of a "semi-institutional" framework "compartmentalizing the economy into industries." Industries would be distinguished by the type of production and the timing of development.[2] Okimoto writes in a similar vein of the "self-contained specialization" of firms into industries, which facilitates ties between state and industry. The cohesive identity of firms within a specific industry simplifies the state's task of developing industry-specific policies. The identity of firms within industries over time fosters the enduring institutional ties between state and industry necessary for

2. Murakami Yasusuke, "The Japanese Model of Political Economy," in *The Political Economy of Japan*, ed. Yamamura Kozo and Yasuba Yasukichi, vol. 1, *The Domestic Transformation* (Stanford: Stanford University Press, 1987), 50.

effective policy enforcement.[3] Identification of individual firms with a specific industry affects relations with labor as well as with the state. Industry labor associations such as the Zensen have come to focus their efforts within industries by first, for example, establishing wage guidelines across the industry, then monitoring negotiations between enterprise unions and individual firms.

The textile industry includes a variety of upstream and midstream sectors, beginning with production of synthetic filament and fiber; and continuing with the spinning of natural, synthetic, and blended yarns; then the weaving or knitting of fabrics; and finally the dyeing and finishing of cloth. Some of Japan's Asian competitors have integrated production across these sectors within a single firm, but in Japan the production process remains largely dispersed across specialized firms, with the companies varying widely in scale even within these sectors. For instance, the synthetics sector of the Japanese industry employed 25,800 workers in 1988 in 108 establishments, and the spinning sector listed 69,000 workers in 816 factories and smaller mills. Yet extensive concentration in synthetics, and to a lesser extent even in spinning stands in sharp contrast to the fragmentation of the rest of the industry. The weaving sector alone employed 148,000 workers in some 32,000 establishments, and knitting 195,000 workers in 17,000 establishments.[4] We also find a huge downstream sector of garment making, wholesaling, and retailing.[5] Restructuring efforts were devoted mainly to the upstream and midstream sectors of the industry, with separate programs designed for the few larger firms on the one hand, and for the large number of medium and smaller enterprises on the other. In general, the larger firms fared better than the smaller firms in the restructuring. The makers of synthetic fiber fared better than the multifiber spinners because of the preexisting concentration of firms in the synthetics sector, the wider possibilities for product specialization in synthetics, and continued strong demand for synthetic fiber and filament.

3. Daniel I. Okimoto, *Between MITI and the Market: Japanese Industrial Policy for High Technology* (Stanford: Stanford University Press, 1989), 125–26.

4. JCFA, *Sen'i handoboku 1992* (Textile handbook 1992) (Tokyo: Sen'i Sōgō Kenkyūkai, 1991), 52. The term "establishment" is a translation of *jigyōsho*. The huge weaving sector falls between spinning and garments on the production line. Ron Dore has written extensively of the weaving sector in Japan. See for instance his *Flexible Rigidities: Industrial Policy and Structural Adjustment in the Japanese Economy, 1970–1980* (Stanford: Stanford University Press, 1986).

5. In 1988 garment making employed 615,268 workers in 50,000 establishments.

Textiles have been among Japan's leading industries since the late nineteenth century, and most of today's major firms boast of pedigrees dating from the turn of the century. The historical record of the firms and employers' associations, and the statistical profile of production and trade chronicle continuity and change (1) within the industry, (2) in the position of the textile industry in the wider domestic economy, and (3) in international markets. The integration and scale of the firms, their products, and their function within the industry as dedicated spinners, integrated textile firms, or diversified enterprises helps locate firms within an industry. Major changes within the industry include product specialization within textiles on the one hand, and diversification out of textile production on the other. The changes in turn affect the role of capital, labor, and the state in the restructuring of the industry. The years have also brought change to the industry's position within the wider domestic economy. Initially a pioneer industry, later a leader in import-substitution and exports, and recently a declining industry, the changes across the years have influenced interest bargaining among capital, labor, and the state.

The international market of textile production and consumption is yet a third context for understanding change within Japan's textile industry. Changes are apparent across the century in Japan's place within an international division of labor for textile production, and in Japan's place within an international textile trade network. The textile industry has depended on foreign sources of raw cotton and until recently, been a net exporter of silk and cotton-yarn fabrics. Again the direction, scale, and balance of trade have helped shape the interests and actions of capital, labor, and the state in the bargaining over adjustment policy. World War II proved to be a watershed in the development of interest group ties within the industry, and so we compare here the prewar and postwar roles of capital, labor, and the state evident in patterns of change within Japan's textile industry, in the place of the industry in the wider local economy, and in the changing position of Japan's textile industry in world markets.

Prewar Years

A review of the prewar history of textile manufacture brings to the fore two features of the early industry: the concentration of capital and manufacturing capacity in large firms, and collective efforts among the firms to balance supply and demand. Joint-stock firms such as Kanegafuchi, Toyo Spinning, and Nisshin Spinning expanded rapidly, often by purchasing smaller mills; by 1907 the top ten firms accounted for more than three-quarters of all spinning capacity.[6] Firms continued to consolidate, but this concentration within the industry did not also result in concentration within a specific region. On the contrary, the initial regional diversity of the mills persisted, which permitted the firms access to a variety of labor and consumer markets.[7]

A further feature of collective efforts among the firms was most evident in the formation and activities of the All-Japan Spinners' Federation (Dainippon Bōseki Rengōkai) from 1882. Three arrangements between spinners and their suppliers and customers helped sustain solidarity within this legally "voluntary" federation. For instance, no spinner could sell yarn to the Federation of Japan Cotton Yarn Merchants Association or import foreign cotton from the trading houses that belonged to the Japan Cotton Traders' Association, unless a member of the Spinners' Federation. And finally, member firms at the federation could receive a freight rebate on all cotton imported from India. Such advantages made membership in the federation a necessity for successful, large-scale enterprise in textile manufacture. But if membership was de jure voluntary and de facto compulsory, there was little doubt about the federation's monopoly of representation for the spinners or about its dedication to balance supply and demand to ensure the profitability of the firms. A former Toyobo executive, Seki Keizo, described the federation as "the medium used to control the production of cotton yarn in Japan."[8]

6. See, for instance, Nisshin Bōseki Kabushiki Kaisha, *Nisshin Bōseki rokujūnen shi* (A sixty-year history of Nisshin Spinning) (Tokyo: Dainihon Printing, 1969), 29–30. Kuwahara Tetsuya, "The Establishment of Oligopoly in the Japanese Cotton-Spinning Industry and the Business Strategies of Latecomers: The Case of Naigaiwata and Co., Ltd.," in *Japanese Yearbook on Business History: 1986*, ed. Nakagawa Keiichiro and Morikawa Hidemasa (Tokyo: Japan Business History Institute, 1986), 110.

7. Yonekawa Shinichi, "The Growth of Cotton Spinning Firms and Vertical Integration," *Hitotsubashi Journal of Commerce and Management* 14 (October 1979): 1–14.

8. *A Study of the Japanese Cotton Industry: Its Past, Present, and Future* (Tokyo: Tokyo University Press, 1954), 63. See also JSA, *Bōkyō hyakunenshi* (A one-hundred-year history of the Spinners' Association) (Osaka: Nihon Bōseki Kyōkai, 1982); Yamazawa Ippei, "Catching-Up Product Cycle Development in the Textile Industry," in *Economic Development and International Trade—The Japanese Model*, ed. Yamazawa (Honolulu: East-West Center, Resource Systems Institute, 1990), 77.

The Spinners' Federation curtailed production to maintain the price of domestic cotton yarn on eleven different occasions before World War II by either reducing the hours of operation or the number of spindles in operation. In the latter case, the federation would register operable machinery and then seal working parts of the machines to be taken out of production. A U.S. government agency surveyed the industry soon after the war and found approximately nine million operating spindles in 1937, the high point of prewar capacity, with nearly two million more (i.e., 18 percent of total capacity) sealed to avoid overproduction.[9] What convinced manufacturers to set aside nearly one-fifth of their machinery? Despite conflicts in the federation regarding production cutbacks and recruitment of skilled workers from rival firms, a common interest in procurement of raw cotton at favorable prices kept the organization together. Kagotani Naoto has written in detail of the compromises reached among the various large firms in the early cartels, and of their strategies to gain the compliance of medium and smaller spinners.[10] Others have written of how early curtailment operations successfully maintained the profits of the firms.[11] And one scholar, Kikkawa Takeo, argued recently that such cartels promoted competition: "On the whole, cartel organizations in Japan could mitigate cutthroat competition and falls of prices during depressions. As a result, member companies of cartel organizations were able not only to evade bankruptcy, but also to reduce loss of managerial resources such as discharge of skilled workers to some degree. It enabled member companies to adopt a growth-oriented strategy based on long-term prospects."[12] Whatever the motive or impetus, capacity reductions apparently worked.

9. Fred Taylor, *The Textile Mission to Japan* (Washington, D.C.: U.S. Government Printing Office, 1946), 1.

10. "Dainippon Bōseki Rengōkai" (The Federation of Japanese Spinners), in *Ryōtai senkanki Nihon no karuteru* (Japan's cartels in the interwar years), ed. Hashimoto Jurō and Takeda Haruhito (Tokyo: Ochanomizu Shoten, 1985), 363–410.

11. For example, Takamura Naosuke, *Nihon bōseki gyōshi josetsu* (An introduction to the history of Japan's spinning industry) (Tokyo: Kōshobō, 1971), 3: 168–75.

12. "Functions of Japanese Trade Associations before World War I: The Case of Cartel Organizations," in *Trade Associations in Business History*, ed. Yamazaki Hiroaki and Miyamoto Matao (Tokyo: University of Tokyo Press, 1988), 86. See also Miyamoto Matao, "Concluding Remarks," 309–10 in the same collection.

Otsuka Keijiro discusses how the federation also ensured that member firms shared technological information, which spurred rapid improvements in technology across the industry.[13] The Spinners' Federation also worked to control labor mobility among mills. Early mills competed to recruit workers from a small pool of machine technicians. Gary Saxonhouse explains the distinctive composition of labor in the textile industry: short-term rather than permanent, and female rather than male. The mills hired large numbers of rural teenage girls who would depart the firm after a few years to return home and marry,[14] but he also highlights the competition among firms for more skilled workers, where experienced technicians were offered heavy inducements to switch allegiance. John Lynch reports that "the Federation was granted enforcement powers by its members, which included the right to hear complaints, investigate charges, and impose rather strong penalties on those found guilty of raiding or poaching from other members."[15] Saxonhouse found the federation ultimately unsuccessful in discouraging interfirm mobility among experienced workers, but adds that there is no evidence of a separate, labor organization among these workers to improve their position with capital.[16]

If we look to capital in the prewar years, we see concentration and cooperation, but always within limits. We find concentration among the early rayon and acetate producers, and to a lesser extent among the spinners of cotton, but also see stability across the period in the hierarchy of larger firms, despite the existence of a large number of medium-size and smaller mills. Apart from the temporary amalgama-

13. Otsuka Keijiro, Gustav Ranis, and Gary Saxonhouse, *Comparative Technological Choice in Development: The Indian and Japanese Cotton Textile Industries* (London: Macmillan, 1988), 72.

14. For a recent study of female textile workers in the prewar years, see Patricia Tsurumi, *Factory Girls: Women in the Thread Mills of Meiji Japan* (Princeton: Princeton University Press, 1990).

15. *Toward an Orderly Market* (Tokyo: Sophia University and Charles Tuttle, 1968), 46.

16. Gary R. Saxonhouse, "Country Girls and Communication among Competitors in the Japanese Cotton-Spinning Industry," in *Japanese Industrialization and Its Social Consequences*, ed. Hugh Patrick (Berkeley and Los Angeles: University of California Press, 1976), 122–23. Walter Goldfrank has linked labor to state corporatist efforts in the alliance of state and capital in the interwar textile industries of Japan and Italy. See his "Silk and Steel: Italy and Japan between the Two World Wars," in *Comparative Social Research*, ed. Richard Tomasson (Greenwich, Conn.: JAI Press, 1981), 297–316.

tion of the industry during World War II, the spinning industry never achieved the horizontal integration found in such oligopolies as the steel industry or the early chemical fiber industry.[17] Concentration was further limited by horizontal, as opposed to vertical integration. Spinners often contracted out weaving, dyeing, and finishing to avoid the expense of integrated production and left the more complex marketing tasks to wholesalers. Also, the All-Japan Spinners' Federation permitted occasional production cartels among the producers to stabilize prices and fostered an exchange of information to promote technological advances, but found itself less capable of convincing members to sacrifice a competitive edge by limiting the mobility of skilled workers. What is striking in this review of limited concentration and cooperation, however, is the prominence and autonomy of capital in the industry from the early years. Initiative for formation of an employers' association (such as the All-Japan Spinners' Federation) came from capital, not from the state. This was not corporatism from above. The firms themselves established the federation to facilitate market relations and lobby the government on trade and tax issues; the federation was not imposed by a directive state on a recalcitrant industry to monitor and control capital.

Looking at the position of labor and the state within the prewar industry, we find labor organized within enterprises rather than across the industry. The Japanese state did mandate a national labor organization during the war years, but only to head off any labor problems that might hamper wartime production. Still, there was a positive side to state involvement in labor questions. Sheldon Garon concludes that "throughout the twentieth century, the Japanese state has played a major role [in labor issues], be it in the form of labor union legislation, the Industrial Patriotic movement, or suppressive red purges." William Wray reports that although labor lacked the peak organizations of cap-

---

17. The reorganization followed the established pattern of concentration in the industry: midsize mills were placed under the administration of the prewar moguls in fourteen units of 500,000 spindles each. Moguls at the time included what we know today as Toyobo, Kanebo, Unitika (i.e., Dai Nippon), Daiwabo, Kurabo, Fujibo, Nisshinbo, Nittobo, and Shikibo (i.e., Fukushima and Asahi).

Among the synthetic-fiber producers, Toray and Teijin began as rayon producers. Teijin traces its roots to Teikoku Rayon, founded in 1918, and Toray to Toyo Rayon, founded in 1926. For a review of the prewar chemical industry, see Barbara Molony, *Technology and Investment in the Prewar Japanese Chemical Industry* (Cambridge: Harvard University Press, 1990).

ital, even at this early stage there was close cooperation between state and labor to ensure some place for labor in negotiations with capital.[18]

The question here is not whether there were ties between state and labor, but how such ties affected interest mediation. Did the state recognize and support the distinct interests of labor, or did the state simply coopt labor to ensure the interests of capital? Pempel and Tsunekawa argue that labor was simply incorporated by the state, evidence of corporatism from above, and that labor never "commanded autonomous legitimacy and recognition."[19] We will look below to labor's status in the postwar years, and suggest here only that disagreement over labor's prewar status can be traced in part to the inconsistent role of the state. Garon and others document inconsistencies in state labor policy. The picture is not clarified by the state's apparent inability to establish a consistent industrial policy. The spinners, for instance, pursued their occasional cartels without sanction from the state and won concessions on tariffs for imports of raw materials. Indeed, the state proved more supportive than directive until the war years, when it tried to manage the industry through the industry control associations. The large firms cooperated with these state organizations on their own terms because the state bureaucracy still had to rely on industry expertise in management and technology. Textile producers quickly recovered their independence following the war.[20]

Turning from the industry itself to its position in the domestic economy, we find the textile sector a major employer and exporter in Japan's early industrialization.[21] A labor-intensive industry in a devel-

18. Sheldon Garon, *The State and Labor in Modern Japan* (Berkeley and Los Angeles: University of California Press, 1987), 248; William Wray, *Managing Industrial Enterprise: Cases from Japan's Prewar Experience* (Cambridge: Harvard University Press, 1989), 77–78.

19. T. J. Pempel and Tsunekawa Keiichi, "Corporatism without Labor," The Japanese Anomaly," in *Trends toward Corporatist Intermediation*, ed. Philippe C. Schmitter and Gerhard Lehmbruch (Beverly Hills, Calif.: Sage, 1979), 253.

20. On the role of capital in the industry control associations, see Richard Rice, "Economic Mobilization in Wartime Japan: Business, Bureaucracy, and Military in Conflict," *Journal of Asian Studies* 38 (August 1979): 659–701, and Peter Duus, "The Reaction of Japanese Big Business to a State-Controlled Economy in the 1930s," *Scienze Economiche e Commerciali* 31 (September 1984): 819–31.

21. "In 1926 nearly 40% of Japan's manufacturing plant was devoted to textiles, producing 45% of her industrial output, employing 53% of her industrial labor force, and supplying two-thirds of the nation's total exports." I. M. Destler, Haruhiro Fukui, and Hideo Sato, *The Textile Wrangle* (Ithaca: Cornell University Press, 1979), 28.

oping economy, the mills substituted local textile products for imports and then set about developing export markets. The growing scale of Japanese textile exports quickly turned the industry into a net earner of foreign exchange, despite reliance on imports of raw cotton from India and the United States. In addition to their leadership in early industrialization, these pioneer joint-stock firms also came to rely on a network of ties with other industries, which sustained further development of production and marketing. Machinery firms such as Howa Heavy Industries, Osaka Machinery Works, and Toyoda Automatic Loom Works were churning out 1.5 million spindles and 20,000 looms annually by 1937.[22] Toyoda had introduced the famous Toyoda automatic loom by 1926 and began exporting the loom to Britain and the United States three years later.[23]

Supported by a local machinery industry for production, the mills found support on the marketing side from both trading firms and shippers. Rapid expansion of maritime transport routes to China, Korea, and Taiwan by such state-supported shippers as NYK (Nippon Yūsen Kaisha) and OSK (Osaka Shosen Kaisha) afforded the mills cheap and reliable access to Asian textile markets. Mills also came to rely on a more uniquely Japanese system of external trade through a network of trading houses, rather than through their own in-house trade departments.[24] The importing of raw cotton was a major factor in the early operations of most of today's big nine general trading companies (sōgō sōsha).[25] Traders established offices in India, the United States, and elsewhere; arranged stocks and shipping; and even provided financing for the mills. Mills in turn could marshal their resources for production rather than marketing, and still find protection from dramatic fluctuations in the price of raw materials despite the absence of a local cotton crop. Moser estimates that Mitsui Bussan, Toyo Menka, and Gosho imported 70 to 80 percent of the raw cotton used by the Japanese mills in the late

22. Taylor, *Textile Mission*, 35.
23. Charles K. Moser, *The Cotton Textile Industry of Far Eastern Countries* (Boston: Pepperell Manufacturing, 1930), 30–32.
24. Takamura Naosuke has written in detail of the early ties between spinners and trading firms. See his *Nihon bōseki gyōshi josetsu*, 1:278–88.
25. Uchida Katsutoshi, "Trading Firms of Japan: The Process of Their Formation and Activities before the Pacific War," *Bulletin of the University of Osaka Prefecture*, ser. d (Sciences of Economy, Commerce, and Law), 2 (1958): 97–107.

1920s.[26] Traders would buy cotton in advance on the New York Futures Market, and then sell that cotton at a later date to a local mill. The traders could offer the cotton at prices a bit below the New York market price because they had already made a profit on the margin of prices between the time they purchased the futures and the time they actually sold the cotton to the mills. Mills found benefit in the combination of the functions of trader and speculator developing at the trading houses.

Ties between mills and trading houses gave Japanese textile firms an advantage over foreign competitors in procuring machinery and raw cotton. But cooperation in long-term business relations did not preclude competition. Mills demanded the best prices and utmost reliability from their traditional suppliers. Earnings from strong sales during World War I permitted mills to purchase raw cotton from the trading houses outright, avoiding the costs of borrowing. Financial autonomy also permitted independence from the prewar industrial combines. Early development of an independent industry allowed the mills to avoid zaibatsu controls.[27] Independent ties with the machinery industry and traders in general fostered growth of the industry. Firms in Japan reported a total of 11.5 million spindles in 1939, second in scale only to the industries of the United States and Britain.[28] Mills spun 870,000 tons of cotton yarn in the peak year of 1937 and 5.7 billion square meters of cotton and hemp cloth.[29] The prominence of the industry in the domestic economy also affected the relationship between labor and capital. Expansion provided more opportunities for employment, but not necessarily for industry-wide organization among labor. Early Meiji state support of shipping and machinery industries indirectly supported the textile industry, but the mills developed largely without direct state sup-

26. Moser, *Cotton Textile Industry*, 35; Seki, *Japanese Cotton Industry*, 124. Takamura compared contracts between the big three spinners at the turn of the century with the emerging trading houses and found great similarities in the protections afforded the spinners. *Nihon bōseki gyōshi josetsu*, 3: 128–40.

27. Nakagawa Keiichiro, "Government and Business in Japan: A Comparative Approach," in *Government and Business*, ed. Nakagawa (Tokyo: Tokyo University Press, 1980), 38.

28. Japan Cotton Traders' Association (Nihon Menka Kyōkai), *Japan Cotton Statistics, 1953* (Osaka: Japan Cotton Traders' Association, 1953), 92–93.

29. MITI, *Sen'i tōkei nempō 1987* (Yearbook of textile statistics, 1987) (Tokyo: MITI, 1988), 40, 43.

port. Expansion at the turn of the century strengthened the independence of the industry from the state.

Since the end of the nineteenth century, textile producers have also
participated in an international network of production and consumption. Otsuka and his colleagues attributed the quick turnaround from
textile importer to textile exporter between 1880 and 1900 to two factors: the shift from mule to ring spinning technology, and the development of cotton-mixing techniques through the transfer and adaptation
of Western technology.[30] One might add the need to gain foreign
exchange for the purchase of imported raw materials, reliable and cheap
means of transport, and access to nearby markets in China and Korea.
Japan's rise as a regional power following the Sino-Japanese and Russo-
Japanese wars assured access to regional markets and protected sea-
lanes. Annexation of Taiwan in 1895 and Korea in 1910, and then
expansion into Manchuria and mainland China in the 1930s opened up
areas for trade and offshore production.[31] Textiles accounted for about
50 percent of the value of the nation's annual exports from 1868
through World War I, jumped to 72 percent in 1926, and was still above
40 percent as late as 1940 as industry shifted to war production.[32]

Japanese political and military hegemony within northeast Asia,
together with government support for colonization in Taiwan, Korea,
and later Manchuria, spurred the initial advance of the textile firms
abroad. Kanegafuchi, Toyo Spinning, Dai Nippon (later Unitika), and
others established spinning plants in Korea and in China, taking
advantage of cheaper labor and energy costs, as well as reduced transportation costs. Operation of a spinning mill abroad would demand
an initial capital investment, technical know-how, and machinery, as
well as a supply of raw cotton.[33] Kanebo, Toyobo, and Dai Nippon

---

30. Otsuka, Ranis, and Saxonhouse, *Comparative Technological Choice*, 72.

31. For an introduction to Japanese economic expansion in Korea before the war,
see my *Colonial Origins of Korean Enterprise, 1910–1945* (Cambridge: Cambridge
University Press, 1990). Carter Eckert has written in detail of the ties of a leading
Korean textile firm with Japanese trading and machinery firms. See his *Offspring of
Empire* (Seattle: University of Washington Press, 1991).

32. Yoshioka Masayuki, *Sen'i* (Textiles) (Tokyo: Nihon Keizai Shimbunsha, 1986), 18.

33. Tōyō Bōseki Kabushiki Kaisha, *Tōyōbō* (Osaka: Toppan Printing, 1986), 307.
For a study of early Japanese investment abroad in textiles, see Kuwahara Tetsuya,
"The Business Strategy of Japanese Cotton Spinners: Overseas Operations 1890 to
1931," in *The Textile Industry and Its Business Climate*, ed. Okochi Akio and
Yonekawa Shinichi (Tokyo: University of Tokyo Press, 1982), 139–65.

had accumulated the necessary capital and expertise and, again, could rely on the textile-machinery industry and the trading houses for technology and raw materials. The firms could take advantage of the security of Japanese political administration in colonial dependencies to ensure the stability of their investment, safely importing Toyoda or other machinery and posting their own technicians to train and supervise local workers. But a spinning mill was one thing, a rayon plant quite another. Establishing a rayon plant involved far more capital, more sophisticated machinery and technical expertise, and more expensive basic materials than would be necessary for a cotton mill. Yet Kanebo's plant in P'yongyang, Korea, was turning out 27 tons of nylon staple, and 10 tons of rayon yarn daily by the 1940s, and Toyobo was producing 5 tons daily of rayon at their plant in Antung, Manchuria. Dai Nippon was producing 24 tons of rayon yarn and 5 tons of staple daily at their plant in Korea.[34]

Investing in the colonies strengthened the hand of capital; firms expanded their production bases and gained expertise in offshore production and marketing. Labor did not profit by colonial investments; on the contrary capital established an early precedent for avoiding rising labor costs and stagnant demand by expanding production abroad. But the real significance of colonial investment lay less in how it diluted labor's leverage with capital than in how it encouraged close cooperation between state and capital. The state made available various indirect supports for private capital investment, without at the same time imposing state direction on the textile industry within Japan. The state helped establish wider directions for industry by encouraging offshore production and rewarding cooperating mills with their support.[35] Meanwhile, the industry retained its independence at least on the home islands, yet could profit from the mills abroad. State and capital found a coincidence of public and private interest in colonial economic expansion.

The prewar years shed light on the origins of a neocorporatist coalition of state and capital in the textile industry and on the prominence of the All-Japan Spinners' Federation. This organization played a

34. Wickliffe H. Rose, *The Rayon and Synthetic Fiber Industry of Japan* (New York: Textile Research Institute, 1946), 8.
35. Concerning incentives offered in the colony of Korea, see my article "The Keishō and the Korean Business Elite," *Journal of Asian Studies* 48 (May 1989): 310–23.

mediating role between state and industry, as well as a coordinating role within the industry. What is significant here is the independent organization of the federation. It was formed without state direction and maintained its independence from the state in representing the interests of capital. Significant also are the patterns of cooperation yet dissent within the organization, with divergences apparent in highly contested areas of competition such as the hiring of skilled labor. Cooperation suggests an "industry interest." Dissent highlights a diversity of firm "interests" at odds with industry-wide efforts at cooperation. One could also point to the subordinate position of labor in the prewar years, and their industry-wide organization under the state only during the war, and then only to ensure productivity and prevent unrest. The review also highlights the prominent role of international ties from the outset, although the industry retained local and regional roots with mills spread across the country. The curious role of the trading houses in both importing raw cotton and exporting yarn and fabrics perhaps diluted the influence of international markets on the producers themselves. Insulated from the volatile raw cotton markets, for instance, the mills were protected from large price fluctuations. The same insulation discouraged in-house expertise in gathering and evaluating market information from abroad. Significant too was the inconsistency in the state's role in the industry, as it moved quickly from its newfound status as a sovereign, independent state at the turn of the century fostering incipient industrialization, to the role of colonizer fostering investment in the empire, and finally to a military state with special powers after 1940 to mobilize the private sector for war. Rapidly changing domestic goals and international priorities in the turbulent years of empire and war precluded a more consistent role in supporting and perhaps directing the industry, yet at no point in the prewar years did the state seriously constrain the enterprise of the textile mills on behalf of the public interest. On the contrary, the state considered employment, production, and export earnings as quite consistent with the commonweal.

Postwar Years

World War II destroyed many of the domestic mills and deprived Japanese firms of their mills abroad. In the spinning industry only about 10 percent of prewar production capacity survived the war.[36] Funding for equipment from the Japanese government and U.S. aid for the purchase of American raw cotton contributed to the reconstruction of the industry over the next ten years. Both wartime consolidation and subsequent reconstruction further concentrated the industry, as plant funding and foreign aid for raw materials following the war were devoted to recovery of productive capacity, rather than to dispersal of assets. This further encouraged concentration in the immediate rebuilding process, although the so-called new spinners also expanded in these years. The Spinners' Association reported seventy-four textile firms with 212 mills in 1936. The association reported one hundred and thirty firms in 1955, but still with only some 200 mills.[37] The industry quickly restored enough capacity to serve domestic needs, and then expanded to supply the U.N. forces during the Korean War (1950–53). An armistice in the Korean War, ceilings on exports to the United States, and a stagnant local market contributed to a recession in the industry by 1956.[38]

The next decade brought new challenges. Asian competitors eroded Japan's share of export markets in Korea, Taiwan, and Southeast Asia. Oil shocks and yen revaluation against the U.S. dollar in the 1970s raised the cost of production and the effective cost of exported goods. Finally, the rapid growth of imports to Japan's domestic markets in the 1980s led to a succession of government-directed programs to control and upgrade capacity in textile production. "Decline" denotes the loss of international competitiveness resulting from comparatively high fixed costs in plant and labor, inefficient equipment, and weak market demand. The restructuring of the industry can be seen as an effort to halt decline and recover competitiveness. Scrap-and-build programs and the search for new products and new markets are long-term efforts to stem decline. The "recession" in the industry is chiefly the result of a temporary imbalance between

36. Taylor, *Textile Mission*, 1. For a brief overview of the prewar and postwar textile industry, see Young-Il Park and Kym Anderson, "The Rise and Demise of Textiles and Clothing in Economic Development: The Case of Japan," Seminar Paper Series, no. 89-04, Centre for International Economic Studies, University of Adelaide, September 1988.

37. JSA statistics are cited from Iwata Katsuo, *Nihon sen'i sangyō to kokusai kankei* (The international ties of Japan's textile industry) (Tokyo: Hōritsu Bunkasho, 1984), 44.

38. Destler, Fukui, and Sato, *Textile Wrangle*.

supply and demand caused by overcapacity. In other words, local and foreign demand have not grown fast enough to make efficient use of installed capacity. Production cartels address short-term problems of recession.

In 1978 both the synthetic- and natural-fiber industries successfully petitioned for designation as seriously declining or "depressed industries," which gained for them state support for retooling and the permission to form production cartels, but also placed them under closer state supervision.[39] The results of restructuring efforts have been mixed at best. Overall, the upstream and midstream production sectors of the industry have become far more concentrated and capital-intensive. About two-thirds of the spinning firms in business in the late 1950s were either absorbed or closed over the next three decades, with the number of firms falling from a postwar peak of 145 in 1959 to 51 in 1990.[40] Restructuring seems to have been more successful in the more concentrated and vertically integrated synthetics industry than in the more dispersed natural-fibers industry.

Concentration and cooperation continue as hallmarks of capital in the industry, with the now familiar prewar moguls retaining leadership of the main industry organizations. The fledgling prewar synthetic-fiber industry was limited to rayon and acetate production, but advanced into synthetic-fiber production by 1952 with government assistance, and then prospered from the 1960s with the shift in spinning from cotton to multifibers.[41] Firms organized the powerful JCFA in 1948, an organization that would play a major role in the restructuring of the synthetics industry from 1978. Meanwhile, the government had disbanded the All-Japan Spinners' Federation (Bōren) in 1942 in favor of a government-supervised Textile Regulation Organization. Spinners regrouped in 1946 as the JSA and quickly found a role in supervising the allotment of raw cotton imported under U.S.

39. Robert M. Uriu, "The Declining Industries of Japan: Adjustment and Reallocation," *Journal of International Affairs* 38 (Summer 1984): 100.

40. Data for the firms in 1959 found in Iwata, *Nihon sen'i*, 44. The total number of firms in 1990 was calculated from listings in Nihon Sen'i Kyōkai, ed., *Sen'i nenkan 1990* (Textile yearbook, 1990) (Tokyo: Nihon Sen'i Shimbunsha, 1990).

41. Tanaka Minoru, *Nihon gōsei sen'i kōgyōron* (A study of Japan's synthetic-fiber industry) (Tokyo: Miraisha, 1967), 47–68; GAO, *Industrial Policy: Case Studies in the Japanese Experience,* Report to the Chairman, Joint Economic Committee, U.S. Congress, October 20, 1982 (GAO/ID-83-11), 46.

aid programs. The JSA later played a major part in gathering information for negotiations on adjustment among state, labor, and the industry, and in shaping and implementing a consensus among capital on production cartels. During the scrap-and-build programs, the JSA assisted the government in gathering and publishing information on equipment registration and production. Both the JSA and JCFA proved to be de facto compulsory organizations for the industry with a monopoly of representation for their members. Both organizations represented the interests of member firms and at the same time helped implement policy directions shaped by both state and industry.

Despite an overall consensus on the restructuring process, one cannot overlook dissent within the JSA, with production cartels particularly distasteful for member firms pushing to expand their share of the market. As one might expect, the mavericks proved recalcitrant, but according to a former director at the JSA, it was difficult to persuade even the moguls to comply with the production cartels of the 1970s. Problems continue. An executive of a major trading house spoke of an effort by the JSA in 1991 to persuade Tsuzuki Spinning to reduce production to stabilize prices. Tsuzuki balked.[42] How does an industry create a consensus and then accommodate dissent without destroying the consensus? It is this blend of competition and cooperation that must be explained in the efforts of capital within employers' associations to design, direct, and implement planned changes.

Certainly the organization of labor within the industry has changed dramatically since the war, but has it grown more powerful? Most observers would agree that labor still does not command a negotiating position equal to that of capital or the state. One might also ask whether labor's peak organization in the industry has accomplished anything. The Zensen (federation of textile and other unions) was established in July 1946 and exercises a good deal of control over member unions.[43] Today it counts as members nearly all of the 61,000 workers in the cotton spinning industry and 53,000 workers in the chemical and synthetics industry; it establishes wage guidelines for the industry and works with member enterprise unions to negotiate

42. The "voluntary cutback" is reported in J. V. Parker, "Cotton Annual Report, 1991," USDA/FAS Report from American Embassy, Tokyo, May 31, 1991, 2. Nittobo also reported a "voluntary decrease in production of spinning products" across the industry. *Nittobo Annual Report, 1991* (Tokyo: Nittobo, 1992), 15.

43. Dore, *Flexible Rigidities*, 138.

annual wage levels.[44] Industry-wide discussions between committees
from the JSA and the JCFA representing the industry, and committees
from the Zensen representing labor precede negotiations at the enterprise level. The Zensen took a central role in negotiating the reduction
of the textile labor force in the restructuring process, though mainly
on factory-level issues rather than on industry directions. Wray suggests that labor's real leverage against capital is exercised through the
state, whereas Pempel finds that labor has no real leverage because it
has no strong peak organization to represent its interests. I will return
to this question, noting here only that labor has become better organized, and organized labor has become more active, since the war.

The role of the state has also changed across the four decades of
postwar recovery. The years of reconstruction found the state in a
gatekeeper role for the industry, allocating state funding, supervising
the distribution of foreign aid, and authorizing establishment of the
"new spinners." Long-term decline in the industry brought the state
into the center of the restructuring process. The latter role commanded more respect from the industry when the state provided funding to help with retiring older and purchasing newer machinery. More
recently, the government has targeted "strategic" or "sunrise" industries for state support, and withdrawn some support from the declining textile industry. Yet three functions of the state remain important
for the industry: authorization of production cartels and other forms
of cooperation to reduce capacity, support for reeducation and reemployment of workers in the process of reducing employment, and
oversight of trade. The state has adamantly refused to impose import
quotas, leaving the market open to all comers on a competitive basis,
but has come to the aid of the industry in cases of dumping, or unfair
subsidies of foreign industries and their exports. No longer gatekeep-
ers for the industry, state bureaucracies remain to help formulate,
implement, and monitor the transition in the industry.

In the domestic economy, textiles have declined in relative importance, in part because of a reduction in textile production capacity,
but mainly because of growth in Japan's chemical and heavy industries, and more recently in areas of high technology. Textiles

44. Zensen, "Zensen 1991," a report prepared by the International Affairs Bureau
(Tokyo: Zensen, 1991). The English title of the Zensen is "The Japanese Federation of
Textile, Garment, Chemical, Mercantile, Food, and Allied Industries Workers'
Unions."

accounted for 12.7 percent of all industrial employment in 1980 and
still 11.1 percent in 1989. Textiles accounted for 5.6 percent of the
total value of industrial production in 1980 and 4.5 percent in 1989.[45]
Within the industry, however, the midstream cotton-spinning sector
reported nearly a 50 percent decline in employment and a 40 percent
drop in the value of production over the past three decades. The synthetic-fiber industry also reported extensive reduction of employment
but only a 10 percent loss in the value of production. Trade statistics
likewise reflect the changing position of textiles among Japan's export
industries. A surprising contrast is evident in the recent turnaround in
the industry from net exporter and earner of foreign exchange to net
importer and consumer of foreign exchange: the steady growth of
Japanese textile exports has not kept pace with the rapid growth of
imports. The annual value of textile exports grew from $1 billion in
1960 to $7.1 billion by 1990, and Japan remains the world's second
largest exporter of synthetic textiles.[46] Yet the share of textiles in the
total annual value of Japanese exports fell from 30 percent in 1960 to
2.5 percent in 1990. Explosive growth in the total value of imported
yarn, fabric, and finished goods left Japan with a negative trade balance in textiles of about $8 billion annually between 1988 and
1990.[47]

A second feature of textiles within the Japanese economy is the continued reliance of mills on allied industries, particularly trading houses
and machinery producers. Trading houses not only import raw materials, they also serve as an integrator in the industry, purchasing yarn
and fabric from spinners and weavers and farming it out to finishers,
and then to wholesalers for sale at department stores and elsewhere.[48]
Trading houses remain both the main sources of raw materials for the
moguls and mavericks and their best customers. Some products they
purchase outright and sell domestically or abroad; other products they
take on consignment for sale overseas. Equally important for the
industry is the availability of state-of-the-art textile machinery from

45. JCFA, *Sen'i handobuku, 1992* (Textile handbook 1992) (Tokyo: Sen'i Sōgō
Kenkyūkai, 1991), 4–5.

46. Data through 1988 provided in ibid., 222.

47. JCFA, *Sen'i handoboku, 1991* (Textile handbook, 1991) (Tokyo: Sen'i Sōgō
Kenkyūkai, 1990), 70–71. One can find data for 1960 in Iwata, *Nihon sen'i,* 14.

48. Data on imports of raw cotton and on the role of trading houses can be found
in my article "A Corporatist Anomaly: U.S. Cotton Sales to Japan," *Journal of International Studies* (Tokyo, Sophia University), no. 30 (January 1993): 51–71.

Teijin Seki or Toray Industries, and also from Toyoda and Howa. Firms such as Nisshinbo and Tsuzuki Spinning invested heavily in new machinery in the 1970s. Others of the mogul firms also upgraded their machinery, resulting in a far more capital-intensive industry in the 1990s.[49] Some of the machinery is purchased in Switzerland or Germany, but the bulk comes from local suppliers, who remain responsible for maintenance through long-term maintenance contracts. Availability of such technology and technicians from local firms has been an advantage for Japanese mills investing in new equipment.

The positions of state and labor with capital have also been affected by the decline in relative status the textile industry has suffered. For instance, the decline of textiles among Japanese industries has prompted textile investors to look to new areas of investment, which has, in turn, prompted textile firms to diversify out of textiles or take their production abroad. As executives at both spinning firms and trading houses put it: "It is not that our share of the local market is declining, but rather that demand for the goods we can produce in the local market is not expanding." As joint-stock firms reporting to financial institutions as their major shareholders, the moguls are being nudged by the banks to find markets where they can expand sales and profits. Meanwhile, their decision to cut capacity and diversify has deeply affected labor. Unions do not fight the decisions to shift production offshore or diversify, but they do make sure their workers are either well compensated for lost positions, or that they receive comparable benefits and the retraining necessary to move to production of nontextile items within the same firm. Labor enjoys considerable leverage in negotiations over plant closings and considerable support from a state hoping to moderate the impact of declining employment and industrial opportunities on local communities. Confident of state support, labor has won admirable benefits in the restructuring of the industry. Here the state must balance a public interest in employment with the private interests of the firms in effective adjustment for survival. The state finds itself in the odd position of simultaneously encouraging the mills to move out of textiles and looking out for the welfare of labor who must bear the brunt of the transition.

49. Anahara Meiji and Suzuki Hajime, "Factory Automation: A Japanese Perspective," in the ITMF's journal, *East Asia: The Textile Perspective* 12 (1989): 42–55.

Changes in the position of Japan's textile industry within the world market have brought remarkable changes to it. No longer competitive either locally or abroad in lower-count cotton yarn and fabrics, or even cheaper synthetics, the industry has moved to higher-quality products.[50] No longer competitive in markets for mass-produced consumer goods, the industry has turned to specialty items, which can be produced in smaller lots. Production of fashion items under various foreign licenses has brought the industry into a new realm of higher value-added products, which can in turn be exported to the markets of advanced industrialized countries. Coupled with more specialized production of high-quality goods domestically, we find also a shift of lower-count production to mills overseas. A Toyobo executive recently told me of productive capacity was almost evenly divided between domestic mills and foreign affiliates. A review of foreign affiliates of the leading moguls shows that the same may be said of other firms as well. Moguls first invested abroad in the 1960s and 1970s, with subsequent capital investment devoted to expanding production at existing affiliates.[51] Originally aimed at import-substitution and market penetration in the targeted countries, the Indonesian and Malaysian affiliates now export a large share of their production to Japan and elsewhere. Offshore production has enabled the moguls to profit at what they do best—the manufacture and sale of yarn, fabric, and synthetic goods. They provide production and marketing advice and direction to the affiliates, assign Japanese plant and office man-

50. "In the spinning process, there is always a fixed relation between the weight of the original quantity of fiber and the length of the yarn produced from that amount of raw material. This relation indicates the thickness of the yarn. It is determined by the extent of the drawing process and is designated by numbers, which are called the yarn count. The standard for the yarn count in cotton is one pound of fiber drawn out to make 840 yards of yarn; the resultant thickness or size is known as count number 1. If the yarn is drawn out farther, so that 1 pound makes twice 840 yards, it is identified as Ne 2 or 2s (i.e., count number 2). A still finer yarn is a ten count yarn, or one pound of cotton drawn out to 8,400 yards. Yarns up to 20 counts are termed 'coarse' yarns, between 20 counts and 60 counts 'medium' yarns, and above 60 count are 'fine' yarns." Bernard P. Corbman, *Textiles: Fiber to Fabric* (Singapore: McGraw-Hill, 1985), 25–26.

The average count of yarns spun provides one measure of product upgrading across the industry. A survey by the JSA reported a rise in the average count of yarns spun among its member mills from 31.77 in 1981, to 34.84 in 1988. "How Yen Appreciation Changed Japanese Spinners," *JTN*, June 1990, 38.

51. Nisshinbo invested in a plant in Fresno, California, in the United States not long ago. Shikibo has recently opened a new affiliate in Thailand.

agers, and recoup their investment with a share of the affiliate's production. The increasing reliance on offshore production as part of the
restructuring process has also attenuated the ties of Japanese capital
with both the Japanese state and Japanese labor.

Faced with the reality of mill shutdowns and lost jobs, labor has
become more concerned with offshore production and with establishing
union networks across national borders in the textile industry. The
Zensen has tried to organize Asian textile workers, but faces obstacles in
gaining the cooperation of unions in less economically developed areas,
or in socialist societies such as the People's Republic of China. Cross-
national ties among trade unions in advanced countries have reinforced
the position of local labor federations in negotiating issues involving an
international division of labor. Such ties have not yet been effective in
fostering labor's leverage in formation of the offshore investment policies
in the Japanese textile industry. A senior Zensen official reminisced that
in the 1960s his organization did not really consider the implications of
either offshore investment or an international division of labor, confin-
ing themselves at the time rather to local issues at the enterprise level.

The importance of the Japanese state in the structural adjustment of
textiles has been somewhat reduced with the overseas investment of
local firms. Capital finds itself in direct negotiation with state officials
in Indonesia or Malaysia in efforts to gain local state support for
import-substitution or export strategies. Japanese textile firms in developing economies find themselves scrambling to gain a share of the
state-authorized quotas for exports to European and U.S. markets. The
Japanese state played a critical role in fostering offshore investment in
the late 1960s when it eased controls on foreign investment, but retains
a role in offshore investment today only through control of trade policy. In earlier decades, the state had to establish quotas on Japanese textile exports to industrialized nations. Today the state is more concerned
with maintaining an open market for textile imports, despite pressures
from the industry for controls to give some semblance of order to the
dynamics of supply and demand in the local market.

Three conclusions can be drawn from the review of postwar changes
in the roles of capital, labor, and the state in the restructuring of the textile industry. First, we find evidence of neocorporatist strategies of
change in the sharing of information among the firms, the production
cartels, the combination of state and industry funding for scrapping and
retooling, and indirect state support through insurance and education

programs for the retraining and employment of former textile workers. Specifically, we find the JSA and the JCFA playing major roles in representing the interests of capital and implementing planned changes, and in mediating relations between capital and the state. Second, we find a weakening of those same corporatist styles of cooperation between state and capital as the state moves from gatekeeper to monitor in the industry, and as the industry becomes more involved in offshore production. The changing shape of international ties offers a further conclusion, as we find capital now responding to an international division of labor in the textile industry. Capital has responded with continued efforts to export high value-added exports to upscale markets and to expand offshore production. International investment has strengthened the hand of capital against the state and labor. And third, the review of the postwar years suggests a shift in the industry away from identification with textile production. Indeed, if Rodney Clark's description of Japan as a "society of industry" is accurate, how will both capital and labor within firms adjust to a new identity as firms producing for multiple industries, rather than firms within industries?

There have been salient discontinuities among capital, labor, and the state within the textile industry itself and in the place of the industry within the domestic economy. Prominent among these is the rise of the mavericks among the new spinners after the war, and their challenge to the mogul hierarchy. The extensive structural adjustment of the firms in this declining industry also stands out, where diversification beyond textiles has begun to redefine both the status and structure of capital and labor. Together with product specialization in a much wider variety of textiles after World War II, diversification has eroded the earlier, cohesive identity of the moguls and their workers as members of a "textile industry." Other changes are evident among labor. Workers developed a peak organization after the war, the Zensen, which now provides an industry-wide voice for labor in annual negotiation of wages, benefits, and working conditions. In the wider domestic economy, textiles lost their prewar prominence among Japanese industries. Other discontinuities among capital are evident in diversification into a variety of industries. We can also point to labor's effort to enhance political leverage by joining other public and private industrial unions in the Rengo.

We find changes also among capital, labor, and state in the international affairs of the textile industry. Capital has been shaken by the foreign penetration of Japan's domestic market and its reversion to the status of a net importer of textiles. The change has also affected the position of labor, which now must look carefully to the survival of the firm and industry when pressing for wages and benefits. Capital can move investment offshore to benefit from the international division of labor, but workers within Japan have no such alternative. Nonetheless, the Zensen has mounted extensive efforts to shape an Asian coalition of labor unions to monitor trade and foreign investment and establish industry standards for wages and working conditions. We can also point to drastic change in the role of the state in the industry's international affairs, where the former colonial power now finds itself negotiating on a more equal basis with its Asian neighbors regarding trade.

Among continuities before and after World War II, we cited the persistence of concentration and cooperation among capital and specified the limits of each. We also found a prominent role for employers' associations representing the interests of capital and implementing the joint policies of state and capital. A shift toward production abroad and the declining importance of the textile industry within the Japanese economy has diluted but not severed ties between state and capital. Looking across industries within Japan, we find the continued reliance of the textile producers on traders and machinery makers. Trading firms and machinery makers supported offshore production in colonial dependencies and China prior to the war, and the same alliance of industries continues to support the efforts of mogul and maverick firms at home and abroad. Finally, we find continuity in the import of raw materials, and in the emphasis on export of finished goods across prewar and postwar periods. This emphasis persists today with more attention to an international division of labor in production, and also to an international division of productive capacity within transnational textile firms. But offshore production within a colonial territory is quite different from direct foreign investment in the industry of a sovereign nation. Close ties between the colonial administration and the investor distinguish the former, whereas joint investment with a local partner in tandem with support from the local government usually distinguish the latter.

The object of this book is the conflict and mediation of multiple special interests in textiles in the course of a transition through decline to stability, but I have not established that the textile industry has actually effected this transition. Robert Uriu cites the textile industry as an example in which "the government has been largely unable to pursue effective restructuring measures." He infers this failure from a continued decline of textile's share of manufacturing and persistent overproduction despite efforts throughout the JSA and JCFA to reduce capacity. Others such as Yamamura Kozo and Brian Ike see the failure of government efforts to force reduction of capacity in the textile industry as evidence that cartels and government subsidies are simply not effective.[52]

Yamawaki Hideki argued recently that the continued international competitiveness of the industry and moderate penetration of the local market by foreign producers indicate success. He concluded that the Japanese industry today is competitive in capital-intensive production of synthetic fibers and yarns, but less so in natural fibers and rayon. Yamazawa Ippei argued similarly that with better technology, product specialization and the growing domestic demand for higher-quality fabrics, the Japanese textile industry has retained its competitiveness in the transition.[53] Earlier evaluations by the OECD and others found the sector-specific, long-term efforts at modernization and scrapping of capacity quite impressive.[54] One conclusion that can be drawn

---

52. Uriu, "Declining Industries"; Yamamura Kozo, "Success That Soured: Administrative Guidance and Cartels in Japan," in *Policy and Trade Issues of the Japanese Economy: American and Japanese Perspectives*, ed. Yamamura (Tokyo: University of Tokyo Press, 1982), 77–112; Brian Ike, "The Japanese Textile Industry: Structural Adjustment and Government Policy," *Asian Survey* 20 (May 1980): 532–51.

53. Yamawaki Hideki, "International Competition and Domestic Adjustments: The Case of the Japanese Textile Industry," Pacific Economic Paper No. 177, Australia-Japan Research Centre, Australian National University (Canberra), November 1989; Yamazawa Ippei, "Increasing Imports and Structural Adjustment of the Japanese Textile Industry," *Developing Economies* 18 (December 1988): 441–62. Analysts within the industry lamented the slow pace of change through 1989, but have pointed to improved labor productivity more recently, and growth in operating profits. See "Reconstruction Needed in the Cotton Spinning Industry," *JTN*, August 1989, 14; "Restructuring of the Japanese Cotton Spinning Industry and the Prospects," *JTN*, June 1991, 67–70; and "Clear Contrasts between Light and Shade—Interim Business Performance for FY 1991," *JTN*, January 1992, 20–22.

54. An OECD report concluded, "On the whole, the record [of adjustment in Japan] appears to be quite impressive." OECD, 143. Jeffry S. Arapan notes that in addition to conventional problems such as high labor costs and rising imports, Japanese firms also "were faced with a currency whose value was rising sharply, and by major reductions in tariff protection." He concludes, "Despite this situation, the Japanese textile complex actually gained strength and increased control over global textile complex activities." He found the Japanese firms more successful at meeting such challenges than U.S. firms. *The Competitive State of the U.S. Fibers, Textiles, and Apparel Complex: A Study of the Influences of Technology in Determining International Industrial Comparative Advantage* (Washington, D.C.: National Academy Press, 1983), 59.

from the debate is that although overcapacity remains, the moguls at least have moved in tandem with the government to reduce capacity and retain a profitable place in textile manufacturing. This new, more specialized, and smaller role in the industry has also been buttressed by investments outside the industry. Reduced capacity and diversification would appear to be the sole solution for stemming decline in the industry, if it were not for the mavericks.

3   The Individual Firm

Firms make the investments, mobilize the personnel, and produce the goods that define an industry. The interests of one or more firms shape the interests of a sector, such as spinning or synthetic-fiber production, and the interests of several sectors shape the interests of an industry. Since late Meiji the textile industry in Japan has witnessed abundant instances of divergence within mutual cooperation, just as in the decades since the war it has witnessed abundant instances of intense competition within industry-wide, common patterns of adjustment. Specialization in microfibers, for instance, has led to intense competition in developing and marketing very similar products such as peach-skin fabrics and imitation leather. Spinners compete to provide ever softer finishes for their cottons. The competition rages well beyond research and manufacture to marketing ties, particularly evident in the licensing of high-status, foreign brand names, but we find not only competition within a common pattern of adjustment but also divergence from the common pattern among dissident moguls and mavericks.

In Japan the individual company is the center of attention in the political economy. The firm provides a focus for the interests of state, capital, and labor that challenges conventional class analysis. For instance, state bureaucracies organize themselves according to firms within industries. Both bureaucracy and Diet committees then work with the major firms in the industry, as well as with the industry associations. Executives work for the firm, not for the stockholders, and labor looks first to the enterprise, and only then to their craft. As a union official at Nittobo put it, "Workers don't get a job, they join a firm." Any effort to unravel the embedded character of state, capital, and labor in Japan must begin with the firm. Mark Fruin recently suggested that "permeable institutional boundaries" lead to growth through alliance and affiliation in the modern Japanese corporation.[1] Japan's synthetic-fiber producers have found sufficient advantages in such affiliations to establish a "downstream" strategy of integration with textile and clothing companies within and beyond Japan.[2] Interdependence between a major firm and its contractors, among firms within groups or *keiretsu*, and across firms within industry associations often dilutes the clear boundaries of ownership and interest more familiar in the West. I suggest that similar parallels of interdependence exist in Japan between state and capital as well as between labor and management within the firm. When I turn to coordination across firms within sectors and among sectors within industries, I find both common direction, as Fruin does, *and* divergence.

Japan has remained a world leader in textiles throughout this period of restructuring. Only Germany surpassed Japan's total of some 900 million tons of exported manufactured fibers in 1989, 11 percent of the total world exports for that year.[3] Japan ranked among the leading producers of spun cotton yarn through the 1980s, surpassed only by the United States, Pakistan, and China. Moreover, Japan increased its yarn production between 1984 and 1988 by 6 percent; at the same time various European nations either reduced production, or added to far smaller bases of spinning capacity.[4] The larger of the mogul firms in Japan rank among the world's leading textile makers. Annual sales at Toray, Kanebo, Teijin, and Toyobo compare favorably with those of Haci Omer Sabanci of Turkey, Wickes of the United States, Hyosung of Korea, and Coats Viyella of Britain.[5] Mogul firms generate big sales

---

1. Mark Fruin, *The Japanese Enterprise System: Competitive Strategies and Cooperative Structures* (Oxford: Clarendon Press, 1992), 9–10.

2. The major synthetic-fiber producers in the United States and Europe have pursued instead an upstream integration strategy of processing their own chemical raw materials. ILO, *General Report of the Twelfth Session of the Textiles Committee, 1991: Report I* (Geneva: ILO, 1991), 68.

3. Fiber Economics Bureau, "Worldwide Manufactured Fiber Survey," *Fiber Organon 62* (June 1991): table 12. Data for 1973 through 1988 can be found in ILO, *General Report*, 73.

4. ILO, *General Report*, 65.

5. Dodwell Marketing Consultants, *Industrial Groupings in Japan 1990/1991* (Tokyo: Dodwell Marketing, 1991), 29; JCFA, *Sen'i handoboku 1992* (Textile handbook 1992) (Tokyo: Sen'i Sōgō Kenkyūkai, 1992), 208–9. Toray and Teijin still fall behind Dupont and Hoechst.

and big profits.[6] Teijin and Toray are usually cited as the major chemical- and synthetic-fiber producers in Japan, with most of their production and sales dedicated to fibers and textiles. Teijin reported sales of $2.3 billion in 1990 and net profits of $144 million; at the same time Asahi Chemical enjoyed sales of $6.8 billion, with operating profits of $274 million. Although fibers and textiles represent only 17 percent of total sales of the latter firm, that is still nearly $1.1 billion worth of fibers and textiles.

Among the smaller synthetic producers, Mitsubishi Rayon and Kuraray reported sales closer to $2 billion, and Toho Rayon sales of $500 million. Bridging synthetics and natural-fiber production, Kanebo enjoyed sales of $3.7 billion, Toyobo $2.4 billion in 1990, and Unitika about $2 billion, with net profits of $24 million at Kanebo, $44 million at Toyobo, and $22 million at Unitika. Nisshinbo led a second tier of spinning moguls with sales of $1.4 billion. The rest of the big ten spinners registered net sales for 1990 of between $500 million and $1 billion, and profits from $5 to $20 million, except for Daiwabo, which registered a deficit.[7] Among the mavericks, Tsuzuki Spinning reported sales of $920 million in 1990, and Kondo Spinning sales of $400 million. The data indicate survival, and even prosperity for some, as Japan's major textile firms have retained their rank in a global industry. Moguls also remain major firms within the domestic economy as evident in employment, sales, and exports, with Asahi Chemical and Toray ranking among the one hundred largest firms in Japan according to total assets in 1990. Teijin, Kanebo, Toyobo, and Unitika found places among the top two hundred, and Nisshinbo among the top three hundred.[8] Within the textile industry, the major synthetics firms have maintained their leadership since the decade following the war. Spinning moguls such as Toyobo,

6. Figures drawn from annual reports of the individual firms; Toyo Keizai's *Japan Company Handbook, First Section, Spring 1992* (Tokyo: Toyo Keizai, 1992); and quarterly reports of the firms filed with the Okurashō (Finance Ministry) and published as *Yūka shōken hōkokusho sōran* (A compendium of financial reports) (Tokyo: Okurashō, 1991). The period reported is the fiscal year ending 31 March 1991.

7. Daiwabo and Shikibo recorded deficits in 1990, but recovered profitability the next year.

8. Asahi Chemical is ranked 77th, Toray 99th, Teijin 129th, Kanebo 141st, Toyobo 178th, Unitika 189th, and Nisshinbo 267th. Nihon Keizai Shimbunsha, *Nikkei eigyō shihyō 1990* (Nikkei financial analysis 1990) (Tokyo: Nihon Keizai Shimbunsha, 1990), 888–89.

Kanebo, and others have been leaders from the turn of the century, but have been joined recently by two maverick firms from Nagoya.

Continuities across the prewar and postwar history of textile manufacture have given shape and substance to a core identity for the textile industry. Yet potentially divisive interests even among firms within common patterns of specialization, diversification, and offshore production have begun to erode that core identity. Patterns of adjustment included both direction and divergence within the industry. At table 4 I compare a synthetics producer, a comprehensive textile maker, a dissenting mogul spinner, and a maverick spinner. Initially organized by the Mitsui Company as Toyo Rayon in 1926, Toray Industries manufactures synthetic fibers and remains a core company in the Mitsui Group with Mitsui Bussan listed prominently among its suppliers and purchasers and ten insurance firms and banks, many affiliated with the Mitsui group, owning 34 percent of its shares.[9] Toyobo prides itself as a comprehensive textile maker with a pedigree stretching back to 1882 as the Osaka Spinning Company. Expanding into polyester production in the 1960s, Toyobo now spins both natural and synthetic fibers. Established in 1907, Nisshinbo also ranked among the larger spinning firms in prewar Japan. The firm today includes a

Table 4. Selected textile producers: Structure, 1990
(assets expressed in billions of U.S. dollars)

Firm	Employees	Assets	Debt/Total assets
Toray	10,428 (−.25)[a]	8.1 (+1.2)	.25
Toyobo	8,001 (−.39)	4.0 (+.86)	.49
Nisshinbo	6,166 (+.07)	2.6 (+2.7)	.32
Tsuzuki Spinning	5,137 (+.46)	—	—

Sources: Please see the source note to table 1. The ratios of total debt to total assets are taken from *Diamond's Japan Business Directory 1991*, with figures for fiscal 1989. Ratios were computed as interest-bearing debts divided by total assets and notes receivable discounted times 100. Data for Tsuzuki Spinning are taken from *Sen'i fashion nenkan 1992*, 332–33.

[a]Ratios in parenthesis represent comparison with base of fiscal year 1980. Toray, for instance, reported a 25 percent decline in employment across the decade, from 13,895 workers in fiscal 1980 to 10,428 workers in fiscal 1990, and total assets increased 120 percent from a level of $3.6 billion in 1980, to $8.1 billion in 1990.

---

9. Toyo Keizai, '92 *Kigyō keiretsu sōran* (An overview of Keiretsu enterprise, 1992) (Tokyo: Toyo Keizai, 1992), 453.

chemical and synthetic fiber department, but remains a major spinner of natural fiber. Our maverick, Tsuzuki Spinning, is a privately held firm based in Nagoya. Established in 1941 as a weaving firm, Tsuzuki Orimono gained the approval of the Japanese government after the war to begin production as a spinner and assumed its present title of Tsuzuki Spinning in May 1948.

These four firms reflect the character of the textile industry and its ties to other sectors and industries. With the exception of Tsuzuki, the selected firms are all listed in the first section of the Tokyo Stock Exchange, with banks and finance and insurance companies appearing as the major stockholders.[10] These firms in turn use the banking services of and insure with their owners. Close, reliable financial ties with the stockholding banks permits extensive exposure for the highly leveraged textile producers, which in turn strengthens ties with their main banks. As is evident in table 5, credit is a primary concern of the firms, because of their relatively high ratio of debt to total assets. The banks maintain a close watch on the firms since they serve as both creditors and owners, but do not directly intervene in day-to-day business interactions, or even serve on the boards except during periods of crisis.[11] Ties between textile manufacturing and the banks provide

10. In 1991, Toray listed the Dai-ichi Life Insurance, Nippon Life Insurance, and Mitsui Life Insurance among its major owners with each holding about 5 percent of the shares. Sakura Bank and Mitsui Trust Bank hold 4.5 percent and 4 percent of the shares. The major owners provided the largest amounts of long-term credits. *Yūka shōken—Toray 1991*, 52.

Toyobo listed four major owners with 3.7 percent each of the total shares: the Dai-ichi Kangyo Bank, Mitsubishi Bank, Sumitomo Bank, and the Nippon Life Insurance Company. These banks also provided the largest amounts of long-term loans for the firm. *Yūka shōken—Toyobo 1991*, 53.

Nisshinbo Industries listed Fukoku Life Insurance with 8.6 percent, followed by Fuji Bank and Dai-ichi Kangyo Bank with 4.9 percent each, and Yasuda Trust with 4.3 percent of the total shares. The Dai-ichi Kangyo Bank and Fuji Bank served Nisshinbo as its major sources of short-term credits. *Yūka shōken—Nisshin Bōseki 1991*, 51. The sources of long-term credits were not listed. The composition of the Nisshinbo board follows the pattern of internal membership of Nisshinbo executives, except for one auditor from Fukoku Life Insurance.

Tsuzuki Spinning listed the Tokai Bank, the Japan Credit Bank, and the Chuo Trust Bank as its major banks in 1990. Nihon Sen'i Kyōkai, ed., *Sen'i nenkan 1990* (Textile yearbook, 1990) (Tokyo: Nihon Sen'i Shimbunsha, 1990), 333.

11. I asked executives at various textile firms why the banks do not post officers to their boards. The textile executives replied there was no reason, as long as the firms were profitable, and added that the banks already knew of the firms' financial status from records of loans and deposits. Thus the boards of the textile firms are or tend to be career employees with no experience outside the firm.

Table 5. Selected textile producers: Performance, 1990
(in billions of U.S. dollars)

Firm	Sales	Profits
Toray[a]	4.1 (+.10)	.205 (+1.3)
Toyobo	2.4 (+.28)	.044 (+1.6)
Nisshinbo	1.4 (+.25)	.046 (+.51)
Tsuzuki Spinning	0.928 (+.46)	—

Sources: Net sales and net profits reported for fiscal year 1 April 1990–31 March 1991. Data are drawn from unconsolidated financial statements reported in *Japan Company Handbook, First Section, Spring 1992.* Data for fiscal year 1 April 1980–31 March 1981 are drawn from *Kaisha shikihō, 57 nen, shinshun* (Company reports, Spring edition, 1982) and from *Nikkei eigyō shihyō 1982* (Nikkei financial analysis 1982, Spring). Data for Tsuzuki Spinning are from *Sen'i nenkan,* 1982 and 1992.
[a]Toray registered a 10 percent increase in sales between fiscal 1980 and fiscal 1990, rising from $3.7 billion to $4.1 billion. Toray reported a 130 percent increase in profits from $87 million in fiscal 1980 to $205 million in fiscal 1990.

some measure of financial security to the textile producers, but the threat of loan defaults or reduced profits at the textile firms will quickly bring closer scrutiny and intervention from the banks.

As apparent in Table 4, each of the four firms is a major employer. The average age of employees at Toray is forty, at Toyobo thirty-four, and at Nisshinbo twenty-nine. The ages reflect not only longer tenure, but also more extensive education and expertise necessary for production at a synthetics firm or comprehensive textile maker such as Toray or Toyobo. I discuss the products, composition of sales, and foreign investments of the four firms later. In general, however, each of the firms targets both local and foreign markets for the products of its Japan-based plants. Exports accounted for 18 percent of Toray's sales in 1991, with customers in Asian nations absorbing about half of this total.[12] At Toyobo, exports accounted for 14 percent of the firm's sales in 1991, with Hong Kong and the United States as the leading destinations.[13] Nisshinbo reported exports accounted for 11 percent

12. *Yūka shōken—Toray 1991;* Toray Kabushiki Kaisha, *Toray gojūnen shi* (A fifty-year history of Toray) (Tokyo: Dainippon Printing, 1977); Toyo Keizai, *Japan Company Handbook,* 237; Diamond Publishing, *Diamond's Japan Business Directory 1991* (Tokyo: Diamond Publishing, 1991).

13. *Yūka shōken—Toyobo 1991;* Toyo Keizai, *Japan Company Handbook;* Diamond Publishing, *Diamond's Japan Business Directory 1991.*

of the firm's sales in 1991, with Asian nations absorbing about half the total.[14] Targeting of export markets, as well as investment in production abroad, indicates a strong international focus at the firms. Close ties with the major trading houses reinforce the international orientation of the producers and also provide channels for penetrating the local market. Mitsui and Company (Mitsui Bussan) appears prominently among Toray's suppliers and customers. Itohchu and the Mitsubishi Corporation (Mitsubishi Shōji) are listed among Toyobo's main suppliers and customers. Tsuzuki Spinning also lists Itohchu as a major supplier and customer for its products.

The recent histories of the four firms provide case studies of the patterns and results of efforts toward collective change. Industry-wide adjustment efforts in any sector demand some degree of uniformity to give substance to the idea of cooperation in a common direction of change. The JSA offered one formula on uniformity and diversity in the textile adjustment: "The 'Textile Vision' will pursue a path for the textile industry worthy of a fully advanced economy in the sense of a knowledge intensive, highly technological industry. Each of the Japanese spinning companies, along this 'hand of the compass,' will continue to seek for one's most suitable way to enforce one's competitiveness."[15] The ambiguity of the statement is one measure of the difficulty individual, market-oriented firms faced in complying with an industry-wide vision of change. The "hand of the compass" suggested in the consensus between state and industry promoted control of capacity and integration of the production process beyond finishing to apparel-making. This chapter follows firms through three steps in the industry-wide vision: product specialization, diversification out of textiles, and offshore investment.[16]

14. *Yūka shōken—Nisshin Bōseki 1991*; Toyo Keizai, *Japan Company Handbook*; Diamond Publishing, *Diamond's Japan Business Directory 1991*.

15. JSA, "Annual Statistical Review of Cotton and Allied Textile Industries in Japan in 1982 and early 1983," *NBG*, June 1983, 5.

16. The integration of the production process has proved far more complex than anticipated. Upstream firms developed as producers of chemical and synthetic fiber, not as integrated textile firms with fiber-making, spinning, weaving, and finishing capacity. Midstream spinners developed as spinners of yarn with limited capacity for weaving. Spinning moguls have set about integrating midstream spinning and finishing, sometimes within the firm but more often with affiliates. Yet a complex web of subcontracting to weavers, knitters, dyers, and finishers remains in place. The trend now toward production offshore, particularly of labor-intensive knitting and sewing operations, has further complicated efforts toward integrated production within the moguls.

Specialization

Producers in Europe, the United States, and Japan all faced decline in the spinning industry in the 1970s. Mills in Germany, Britain, and the United States opted for mass production of commodity products to boost production, and only the United States firms with a huge domestic market succeeded. Italy and Japan opted rather for specializing in fashions and household and industrial textiles.[17] Why did Japan choose specialization? The rise of integrated textile firms in Asia with lower labor and energy costs in the 1960s posed two threats to the Japanese textile industry. Import-substitution strategies among mills in developing Asian economies threatened to deprive the Japanese firms of a traditional export market for their lower- and medium-count yarns and fabrics. Second, export-oriented strategies in the same developing economies of Asia threatened to bring cheaper textile imports to Japan, forcing the Japanese firms out of a profitable market niche in their own backyard. Competition from Asian nations with lower labor costs and often also a local supply of raw cotton led most of the major spinners in Japan to curtail production of commodity cottons, although we find divergence among the firms in specialization and upgrading of cotton production.

Japanese producers responded to the initial threat with some investment abroad in Brazil and Southeast Asia in the 1960s, but devoted most of their resources to restructuring productive capacity at home. The state worked with the industry to develop a scrap-and-build program from 1964 to upgrade capacity but curtail expansion. This strategy of labor- and energy-saving technology to offset the advantage of lower labor costs among their foreign competitors proved at best a partial solution to changes in an international division of labor. Expansion of local production in developing countries and their expanding share of textile export markets in the 1970s forced Japan to reappraise production plans at home and abroad.

The problem in the 1960s was not yet penetration of Japan's domestic market, but rather loss of traditional export markets for Japan's textile products. Despite such losses, the Japanese textile industry could look to other areas of growth. For instance, mills could

---

17. Geoffrey Shepherd, *Textile-Industry Adjustment in Developed Countries* (London: Trade Policy Research Centre, 1981), 4–5, 21.

supply new demand for more sophisticated textile products at home among increasingly prosperous Japanese consumers and still expand established trade networks in North American and European markets. Market opportunity was not the only advantage keeping Japanese producers in this "declining" industry, for textile producers interested in upgrading could obtain sophisticated spinning and weaving machinery from local machinery makers. Toyota and Howa remain today leading international suppliers of sophisticated spinning and weaving machinery, and affiliates of Teijin (i.e., Teijin Seiki) and Toray (i.e., Toray Engineering) manufacture state-of-the-art machinery for production of synthetics. If the shift from cotton to multifiber production distinguished the 1960s, the 1970s saw product specialization in more sophisticated textiles with higher value-added content. The industry had hit on a viable strategy of adjustment for the depressed industries of both synthetic- and natural-fiber production.[18]

Nonetheless, a sense of crisis swept the industry in the 1960s when the first imported textiles arrived. The chairman of Toyobo conveyed the gravity of the situation by sharply criticizing Japan's spinners for having no vertical integration, that is, for being limited to selling cloth and yarn, which led, he argued, to a preoccupation with physical productivity at the expense of quality, and with mass production at the expense of attention to the consumers' needs.[19] His colleagues had heard the message before. Now it was time to act on it. Government-supported programs for scrapping and upgrading equipment helped textile producers replace older equipment with more versatile and efficient machinery, which permitted the manufacture of multifibers, blended fabrics and yarns, and higher-count cottons. The best evidence of the trend toward improved technology was the shrinking

18. Harold Sloan and Arnold J. Zurcher offered the following definition of *value added*: "For a given enterprise, the market price of a good, less the cost of materials purchased from others and used to fabricate that good. Gross value added includes payments for taxes, interest, rent, profits, reserves for depreciation, and compensation to management and other employees." Sloan and Zurcher, *Dictionary of Economic Terms* (New York: Barnes & Noble, 1970), 459. The amount of value added (*fuka kachi*) is computed from the sum of personnel costs, loans, taxes and fees for licenses, depreciation on machinery, and operating profit. The ratio of value added to sales is computed by dividing the net amount of value added produced by net amount of sales in a given year, and multiplying by 100. Nihon Keizai Shimbunsha, *Nikkei eigyō shihyō 1990*, 12.

19. Taniguchi Toyosaburo, "Man-Made Fibers and the Changing Textile Industry," *NBG*, no. 237 (September 1966): 64–72.

labor force and rising ratio of productivity per worker. The key to specialization is the "proper mill balance" between productivity and flexibility, which demands planning on the one hand, but also the
shopfloor skills of technicians who can calibrate machinery. Frequent
style changes vastly complicate the process of fully utilizing all equipment.[20]

Statistics on annual production show a trend toward blended fabrics and higher-count cottons. Annual volume of natural fibers and
fabric fell 15 percent between 1961 and 1970, 35 percent by 1980,
and finally to about 50 percent in 1990 of the peak volume three
decades earlier. Annual production of pure cotton yarn has fallen
some 26 percent from its peak in 1961, although the production of
much smaller volumes of blended cotton thread has expanded.[21] The
industry still produced 426,000 tons of cotton yarn in 1990, and 2.2
billion square meters of fabric from synthetic fibers.[22] Meanwhile, the
volume of fabrics made from chemical fibers rose 65 percent between
1961 and 1970, held steady for a decade, and then fell about 21 percent by 1990. Annual production of blended synthetic yarns grew
until 1986, and then fell off recently. In 1990, the nation produced
383,000 tons of blended yarn, and 3.3 billion square meters of fabric
from synthetic fibers.[23]

The tendency in the industry toward higher-count cottons and
toward blended yarns and fabrics suggests a common trend, but in
fact the different firms brought out a variety of products as they competed for domestic and foreign market shares. Nisshinbo and Tsuzuki
never made the investment in producing polyester, focusing instead on
natural fibers and on spinning blends of synthetics and natural
fibers.[24] Toyobo produced both synthetic fibers and cotton blends,

20. ILO, *Vocational Training and Retraining in the Textiles Industry* (Geneva: ILO,
1991), 11.

21. A 50 percent decline in production of medium-count (23–44) cotton yarn and a
300 percent decline in the volume of 20-count yarn accounted for the fall in annual volume of production. Yet annual production of fine-count cotton yarn (45 and above)
doubled across the same period, and production of coarse counts (below 20) held
steady. JSA, *Statistics on the Japanese Spinning Industry 1991* (Osaka: Japan Spinners'
Association, 1992), 45.

22. JCFA, *Sen'i handobuku 1992*, 25.

23. JCFA, *Sen'i handobuku, 1992*, 25.

24. Nisshinbo is the major shareholder in one of the synthetics moguls, Toho
Rayon, with 24 percent of the stock.

and Toray specialized in synthetic fibers. Toray initiated production of
nylon in 1951, and their colorfast nylon filament remains a major
product today. The firm later developed Tetron polyester staple fiber
and filament yarns and by mixing polyester filaments of various diam-
eters and shrinkage rates now produces the so-called Shingosen (new
synthetic fabrics).[25]

Toray is a mainline producer of synthetic fibers. Toyobo tries to
bridge synthetics and spinning of natural fibers. Polyester accounts for
nearly half of Toyobo's sales in synthetics, evidence that like Toray,
Toyobo has invested heavily in the new synthetic fibers based on polyester blends. Toyobo reports growth of sales in new products such as
Geena, a polyester filament fabric, and also in spunbonded polyester
for industrial use. They have developed polyurethane elastane fibers,
protein-grafted fibers, and polynosic fibers for use in spun and woven
fabrics. In natural fibers, Toyobo is the largest purchaser of high-quality American Pima cotton in Japan and a major purchaser of Egyptian
high-quality cotton. The firm spins ultra-high-count cotton fabrics
such as Toyobo Isis and Toyobo Supima for specialty markets.[26] A
Toyobo executive explained in the fall of 1991 that fully a third of
their local production of cotton yarn was now above 60 count, that
the company regularly spins up to 250-count yarns, and that their
mills have the capability of spinning up to 300-count cotton yarns. A
four-hundred pound bale of Toyobo's high-count yarn, used in
women's blouses and scarves, could earn a big profit, although
demand remains sluggish. Toyobo's emphasis on specialization in
both synthetics and natural fibers sets it apart from synthetics giants
like Toray and Teijin.

As a dedicated spinner with a chemical- and synthetic-fiber department, Nisshinbo has chiseled out for itself a third path of specialization quite distinct from the paths of Toray or Toyobo. Nisshinbo has
identified itself with a motto of "distinctiveness" in cotton products
and highlights the sophistication and efficiency of its production of
cotton and synthetic fabrics.[27] The firm has succeeded in spinning and

25. Toray has also produced a synthetic suede. See Toray Industries, *Toray: Annual Report 1991* (Tokyo: Toray Industries, 1991) and *Toray Kabushiki Kaisha* (Tokyo: Toray Industries, 1991).

26. Toyobo Company, *Toyobo Annual Report 1991* (Osaka: Toyobo, 1991).

27. JSA, *Statistics on the Japanese Spinning Industry 1968* (Osaka: Japan Spinners' Association, 1969) and *Statistics on the Japanese Spinning Industry 1991*.

weaving "the world's thinnest cotton yarn," a 310-count yarn. As evidence of their commitment to ultrafine yarn production, Nisshinbo has funded research at Texas A&M University in the United States in the use of biotechnology to develop long-fiber cotton strains. They have also tried to advance their manufacturing techniques at the other end of the production process by improving the finishing process to produce more specialized cottons. The transition for Nisshinbo has been from natural fibers to multifiber production on the one hand and to improvements of product and process on the other hand. They note proudly in their annual report, "We concentrated on the top end of the market." Coupling offshore markets with specialization proved successful: "Sales of value added products overseas more than compensated for the decline in the home market."[28]

Yet a fourth path of specialization is evident at the maverick Tsuzuki firm, where they have concentrated on medium-count yarns (23–44). We find a similar pattern at a subsidiary, Shin Nippon Spinning, which specializes in coarser cotton yarns. But unlike its parent firm the affiliate at least opened lines for spinning synthetic fabrics and yarns.[29] The competitive challenge facing Tsuzuki in this market niche is the installation and effective use of labor and energy-saving equipment to offset the lower labor and even energy costs of their offshore competitors. One example of their challenge is the integration of machinery from different countries. A Tsuzuki executive complained, for instance, of the complexity of harmonizing software programs between computer-assisted machinery from Switzerland and Japan. Like Nisshinbo, Tsuzuki boasts of highly efficient production facilities, and of its "TNS" production system.[30] Specialization among the moguls denotes new products for smaller, upscale markets, but Tsuzuki Spinning has remained with the earlier pattern of mass-producing yarn and fabric for a medium-quality market.

28. Nisshinbo Industries, *Nisshinbo: Annual Report 1991* (Tokyo: Nisshinbo, 1991), 6. The reference here is to markets in the United States and Europe: "Our super soft cotton products sold well in western markets, while our top-range cotton-polyester fabrics for pret-a-porter as well as haute couture apparel were particularly well received in Europe."

29. Tsuzuki Spinning owns 55.8 percent of the shares of Shin Nippon. Toyo Keizai, *Japan Company Handbook, Second Section, Spring 1991* (Tokyo: Toyo Keizai, 1991), 125.

30. Tsuzuki Spinning, *Aggressively Tsuzuki* (Nagoya: Tsuzuki, 1991).

One might attribute the different paths of specialization to plant endowment and expertise, but clearly marketing strategies also play a part. Toray could specialize in new synthetic products because of its knowledge of polymer technology. With total assets of $8 billion and a relatively low debt burden compared to its competitors, Toray has the financial clout to make the investment necessary for research and design of the new synthetics. Toyobo cannot match those resources. Toyobo lists total assets of $4 billion and a relatively high debt burden. The firm still can call on considerable assets for research and on a base of technology and expertise for specialization in cotton blends and some new synthetics. Nisshinbo reports total assets of $2.6 billion. It has remained a dedicated spinner of cotton and multifibers and has made the investment in efficient technology to maintain profitability in that area. A choice to remain in a declining market at home and a soft market abroad hardly coincides with the industry vision. The decision to remain in mass production of middle-count cottons runs counter to the industry emphasis on product specialization. How can Tsuzuki hope to compete with other Asian producers in the medium-count market in Japan and abroad? They argue that advantages of technology, know-how, and proximity to the market can offset the differences in labor and energy costs between Tsuzuki and producers from the developing areas.[31]

Diversification

Firms leave a market when there is little prospect of growth, or worse, when they cannot maintain their place in it. The variety of products possible in textile manufacture, and the potential for new consumer and industrial textile markets in Japan and other advanced industrial economies, left Japan's mills agonizing over whether to abandon potentially lucrative textile markets for unknown territory in other product or service lines. The textile industry in Japan had been in decline since the late 1950s, and gained state recognition as

31. In an interview on February 18, 1992, Tsuzuki officials did note that if an Asian producer had access to local cotton prices lower than those in the world cotton market, they would have a big advantage over the Japanese producers. They pointed out that both Pakistan and the People's Republic of China have local supplies and could become serious competitors.

"depressed" in 1978. The law on depressed industries stipulated three criteria for this designation: "a) more than 50% of the industry's firms must be experiencing financial difficulties; b) the industry must be characterized by unusually excessive plant capacity; and c) firms representing two-thirds of the industry must sign a petition seeking designation as structurally depressed."[32] As Yoshioka Masayuki remarked at the time, "The textile industry has declined in terms of amount of product shipments, amount of value added, and number of employees."[33] Both the synthetic-fiber makers and the natural-fiber spinners were granted this designation and government support from 1978 to 1983.

The government's designation of growing industries as "strategic" provides a useful foil for the depressed industries. MITI offers four criteria for "strategic" industries: (1) efficiency of capital investment; (2) comparative advantage in the international division of labor; (3) capability for import-substitution industrialization, and for growth into an export industry; and (4) employment opportunities at affiliated medium and small industries.[34] Yet the opposite of a strategic industry is not a "declining" but a "sunset" industry, for the textile industry although in decline was still expanding its exports, could probably establish a new, productive niche in the international division of labor, and could still provide employment opportunities in subsidiaries or affiliates. Integration and product specialization were options for improving efficiency within the industry, just as scrap-and-build programs and production cartels were options for restructuring and salvaging it. But diversification was a more difficult choice since it took

32. Comptroller General of the United States, *Industrial Policy: Japan's Flexible Approach*, Report to the Chairman, Joint Economic Committee, U.S. Congress, 23 June 1982, GAO/ID-82-32, 68.

33. "Overseas Investment by the Japanese Textile Industry," *Developing Economies* 17 (March 1979): 17. One should note that although value added may have declined in the wider industry, the amount of value added increased among the top eighty firms between 1969 and 1978. The rate of increase between 1969 and 1978 was only slightly lower than the average rate of growth for all manufacturing sectors in the same period. Japan Development Bank, *Handbook of Financial Data of Industries, 1979* (Tokyo: Japan Development Bank, 1979), 46, 62.

34. Iichimura Shinichi, "Japanese Industrial Restructuring Policies, 1945–1979," Center for Southeast Asian Studies, Kyoto University, November 1979, 12; Suzumura Kotaro and Okuno-Fujiwara Masahiro, "Industrial Policy in Japan: Overview and Evaluation," Pacific Economic Paper No. 146, Australia-Japan Research Centre, Australian National University (Canberra), 1987, 25; Bernard Eccleston, *State and Society in Post-War Japan* (Cambridge: Basil Blackwell and Polity Press, 1989), 42.

these firms beyond their home industry into new areas of investment, production, and marketing.

How do textile firms diversify? European or American firms might begin with purchases of stock in a variety of industries. But since most Japanese firms are closely held, in Japan one seldom diversifies through large stock purchases or takeovers. Although some firms have investment portfolios with a variety of stocks, most tend to invest in the very financial institutions that have invested in them. Firms also hold stock in other banks and often in other sectors of the textile industry. The textile firms usually expand into new areas, not through a varied portfolio or takeovers, but by opening a new product line, supported often by a new network of subcontracting companies. Like their European and U.S. counterparts, Japanese firms have been actively pursuing diversification.[35] To branch out from textiles successfully, Japanese firms have found it necessary to establish a base of expertise, gather the requisite financial resources, and time their investment properly. For instance, Kanebo moved into the unrelated area of cosmetics in the 1960s despite the lack of related technology or expertise. Here timing and finances were the key. Few textile firms today could afford the start-up costs of research, production, and marketing necessary to join Kanebo in cosmetics. Kanebo has distinguished itself in the industry with diversification into cosmetics, foods, and pharmaceuticals, but remains basically a textile firm with fashion merchandise, natural- and synthetic-fiber production still accounting for over 50 percent of its sales. By contrast textiles account now for only 17 percent of the sales of Asahi Chemical, which has moved aggressively into nontextile areas such as building materials and housing and maintained production of chemicals and plastics. Kanebo has retained its identity within the industry despite investment in unrelated areas; Asahi Chemical has not.

Textiles still account for better than half of their total sales at the mogul firms. Diversification into areas parallel to textiles is usually preferred, if a profitable market niche can be developed. As the editors of Nisshinbo's annual report declared, "Given the opportunities to apply accumulated technologies in other areas, diversification has

35. United Nations Centre on Transnational Corporations (UNCTC), *Transnational Corporations in the Man-made Fibre, Textile and Clothing Industries* (New York: United Nations, 1987), 64.

long been an important element in the Company's operations."[36] For instance, Unitika and Daiwabo have diversified into textiles for industrial use; Nisshinbo has started making the materials used in brake linings. Firms appear more confident with such parallel diversification because of their base of production expertise and knowledge of markets. In this case polymer technology with its multiple applications has provided the synthetic-fabric makers with an advantage over spinners and their limited technology. Expertise in chemical compounds and some familiarity with related markets has permitted firms to expand their chemical production or move into production of plastics. In contrast, much of the investment for the mogul spinners has gone toward quite new fields, rather than parallel areas. Kurabo is trying to find niches in interior decoration and information systems. Nittobo has long been producing glass fiber, but is now designing the fiber for new applications. Nisshinbo is producing label-printing machines.

A closer look at diversification strategies and results at our four target firms reveals quite a variety of efforts and products. At Toray polyester, nylon, and acrylic fiber accounted for 93 percent of total sales in 1971. Two decades later, textiles and fibers accounted for only 48 percent, plastics and chemicals 26 percent, and housing and engineering 15 percent of sales.[37] Housing and engineering represent sectors of investment quite new to the firm, but Toray could build on the firm's own expertise to move ahead in more diversified chemical and plastics. In contrast, Toyobo has not ranged far beyond its original expertise, but has certainly capitalized on its base of technology and expertise to diversify in parallel areas. Cotton and wool represented 40 percent of Toyobo's sales in 1971, with sales of rayon, polyester, nylon, and acrylic staple accounting for the rest. Two decades later, natural fibers and textile products account for 34 percent of total sales at Toyobo, synthetic fibers and textile products another 34 percent, packaging films 10 percent, and biomedical products 6 percent.[38] The latter data indicate both specialization and diversification. Toyobo has moved from spinning to more integrated textile production, including garment making with subsidiaries and affiliates, but the firm has also

36. *Nisshinbo Annual Report 1991*, 1.

37. Noyes Data Corporation, *Textile Industry of Japan, 1971* (Park Ridge, N.J.: Noyes Data Corporation, 1971), 139; *Toray Annual Report 1991*, 21.

38. Noyes Data, *Textile Industry of Japan, 1971*, 143; *Toyobo Annual Report 1991*, 35.

moved out of textiles into parallel areas such as nylon packaging films.

Nisshinbo has long been active in diverse areas of investment, and now styles itself "Nisshinbo Industries, Inc." The company reported that textiles accounted for about 72 percent of sales two decades ago, with paper, urethane foam, and other products already part of the firm's product mix.[39] Textiles represent 64 percent of total sales today. Textile sales still include yarn and greycloth, but the value of finished goods such as shirt materials, bedding materials, and denim is growing. One interesting example of diversification in the industry is the production of disc brake systems at a textile firm. Nisshinbo has been providing asbestos and other brake lining materials to the automobile industry for some time and has recently obtained licenses from British and German firms to produce disc brake systems and other friction materials. Here, an initial investment in a related or parallel diversification, such as asbestos and brake linings, grew over time into quite unrelated areas, but production of disc brake systems became feasible only after Nisshinbo established a market share in brake linings. Nisshinbo also produces label-printing machines and hopes to divide its sales evenly between textiles and nontextiles by the turn of the century.[40] Tsuzuki Spinning also has invested in service industries in recent years, although since Tsuzuki is privately held, data on its investments and sales are not so easy to come by. The company has moved into biotechnology research, real estate and warehousing, development and management of golf courses, and the hotel and food service industry.[41] Interviews with executives at the firm suggest the outside investment is still considered secondary. Tsuzuki is a spinner.

The review of specialization above indicates market strategies based on the endowment of the firms. Firms specialize and expand according to their base of technology and expertise. The consensus on restructuring hammered out between state and industry provided some resources for upgrading technology or moving out of textile production, but the initiative to exploit those markets remained with the firms. On one level, the consensus depends on a common overall vision and policy of adjustment, but on a more practical level, progress depends on rela-

39. Noyes Data, *Textile Industry of Japan 1971*, 95.
40. *Nisshinbo Annual Report 1991*.
41. *Aggressively Tsuzuki*, 4.

tively stable market shares and at least a minimum of regular profits in a changing competitive environment. Spinners still produce huge amounts of cotton and blended yarns, and huge lots of greycloth. Synthetics firms produce tons of staple and filament yarns. Specialization demands capital investment, and firms that cannot maintain their profits are hard-pressed to convince their banks to fund new investment. Industry-wide cooperation is critical if production cartels are to maintain some balance between supply and demand. Some restraint among those firms seeking to expand market share at the expense of the adjusting firms becomes similarly important.

If cooperation and restraint are important during the vulnerable years of research and development of specialized products, the same concern for the stability and "reasonable profits" is important during times of investment in areas of nontextile production. Diversification demands more than large amounts of capital. It also involves time-consuming planning and preparations for entering new markets and patience while waiting for long-term investments to begin to bear profit. But firms and markets do not stand still. A firm can tolerate low returns in new investment areas only if there are consistent returns in their central area of investment. Mills have to turn a profit in textile production to generate the capital for investment in nontextile areas. Maverick behavior in a period of adjustment can disrupt the very profits necessary for adjustment. We need to explain why the moguls tolerated the expansion of spinning capacity not only at Nisshinbo during these two decades of diversification but also—and especially—at the mavericks like Tsuzuki and Kondo Spinning.

Offshore Production

Product specialization within textiles and diversification into unrelated areas of production or services represent two ends on a continuum of adjustment programs. The former strategy draws a firm deeper into the industry; the latter moves a firm out of the industry. Offshore production falls closer to the middle of such a continuum, for even though it is a shift out of the local industry, it is based on existing technology, as well as in-house production and marketing expertise. Unlike their competitors in Europe and the United States, Japanese firms have made a variety of offshore investments, through a variety of ownership structures. Japanese firms have established production bases in Latin America, Southeast Asia, and Northeast Asia, as well as Europe, whereas their European counterparts have focused investment in economically advanced nations in Europe and the United States. Furthermore, Japanese firms usually go abroad in partnership with trading companies, and with local partners in the target country as well, rather than as sole owners of offshore affiliates.[42]

The textile industry was among the first of the postwar industries in Japan to look abroad for lower labor and production costs.[43] An early advance abroad of the spinning firms can be explained by market dynamics. Lower-cost competitors slowly nudged Japanese mills out of lower-count markets abroad and eventually at home. On the positive side, investment incentives from governments of newly industrializing countries anxious for foreign capital and technology attracted Japanese textile firms. On the negative side, it was clear already in the 1960s that even with upgrading of plants within Japan, the industry's traditional role in the markets of developing countries could only be maintained by shifting production to those nations. Spinners responded by building plants in Brazil, Central America, and Southeast Asia in the 1960s, but the total investment in mills abroad between 1951 and 1969 amounted to only $140 million.

The Japanese government began to liberalize procedures for investment abroad between 1967 and 1970. At the same time governments in Indonesia, Thailand, and Malaysia offered incentives for foreign investment in import-substitution production. The textile industry initially invested in the mills of Southeast Asian nations, establishing affiliates and subsidiaries that were to produce for local markets. Tax benefits and other incentives from the local governments helped defray costs of transferring spinning machinery, and even technology for production of synthetics. Japanese textile investments in Asia alone amounted to $1.8 billion by 1991; almost half of Japanese investment worldwide in offshore textile production, and textile investments account for almost half of total Japanese foreign investment in manufacturing in Asia today.[44] An analyst for *Japan Textile*

---

42. UNCTC, *Transnational Corporations*, 33–39.
43. For a summary of investments in offshore production, see "The Global Network of Japanese Textile Companies," *JTN*, June 1991, 71–80.
44. JCFA, *Sen'i handobuku 1992*, 377; Keizai Koho Center (Japan Institute for Social and Economic Affairs), *Japan 1991* (Tokyo: Keizai Koho Center, 1990), 43.

News described the investments in ASEAN countries as "sub-mill" efforts, with production of intermediate yarns and gray fabrics exported for further finishing, or even brought back to the home firm in Japan for finishing and marketing. Earlier Latin American investments may align more with an "anticipated dividend" strategy, with Japanese firms hoping to earn dividends and commissions from technical guidance. We also find Japanese mills now in the United States and European Community, which suggests a third investment strategy of "capital diversification" into the prosperous industry and markets of another advanced, industrialized economy.[45]

The consensus between industry and the state in Japan on structural adjustment for the depressed natural- and synthetic-fiber sectors emphasized the need for investment abroad, but many of the moguls had been active abroad well before 1978. An upward evaluation of the yen and weak performance of the textile firms over the next decade slowed investment until 1987, when foreign penetration of Japan's import market began to expand dramatically. Japanese firms invested $1.8 billion in textile production facilities abroad in the four years between 1987 and 1990. Synthetic-fiber makers and spinners have followed different strategies of investment abroad, although both groups begin with local partners and their investments often continue as joint ventures. Spinners often ally with a Japanese trading company as well as a local partner; the spinner and trader each hold 25 percent of the stock, and the local partner holds the rest. Yoshioka explains that the spinners lacked the expertise and capital to invest alone. Still another reason for the alliance with traders would be insulation from the risk of investment as the sole foreign partner.[46] Interviews with executives of Japanese affiliates in Indonesia and Thailand confirm that the goal of the spinners has been either to produce for the local market or to develop platforms from which to export to still other countries. It was also clear that the spinners make money in the ventures by selling know-how, machinery, and plant facilities.

Toray has multiple investments in Thailand, Malaysia, and Indonesia, but also in Korea, Taiwan, and the United States, and is now expanding in Britain and on the European continent. In Indonesia, Malaysia, and Thailand it has focused on integrated production from

---

45. "Editorial: Japanese Spinners Invest Abroad Actively," *JTN*, February 1990, 16.
46. "Overseas Investment."

polyester and nylon-filament and cotton spinning to weaving, knitting, and even dyeing, though the more sophisticated technology and expertise necessary for dyeing is still difficult to establish. The further task of integrating production across borders still eludes Toray.[47] In an interview with an executive at Luckytex in Bangkok, of which Toray owns 49 percent, I learned of various joint projects between Tokyo and Bangkok. The parent firm also serves as broker, importing more than half of the raw cotton Luckytex requires, provides information on changes in fashions, and projects demand for various target markets. The Tokyo headquarters plays a major role in planning for changes in production targets, improving facilities, and importing machinery, but Luckytex sets its own production goals and establishes its own marketing plan. Apparently, independence in production and marketing makes Luckytex more efficient in meeting demand abroad.[48] Luckytex exports about two-thirds of its production, the lion's share directly, with only 10 to 15 percent sold to local garment-making firms, which then export finished goods.

Toray has invested offshore to develop a global network of production and marketing. Toyobo went abroad initially to develop a niche in local markets. But with further industrialization in the developing countries, the local governments offered support for firms exporting their production, permitting the joint ventures of Toyobo to turn their energies to markets abroad. Toyobo has joint-venture spinning investments in Brazil and Central America, Thailand, Malaysia, and Indonesia. Executives at Dusit Erawan in Bangkok, a Toyobo joint venture with Marubeni and local investors, introduced me to the organization and goals of the joint venture. I found a clear division of labor among the spinners and trading companies, with Japanese administrators from the home firms holding the managerial positions at Erawan. Toyobo would post plant managers, production managers, and raw-material overseers to the joint ventures for three-year terms, and the trading houses would send their managers to manage finance and trade operations. Toyobo hopes to increase "communication among our many affiliated companies to bolster efficiency in production and marketing," but integration of even the textile production

47. The firm remains committed, however, to "globalization." "Toray Sets New Mid-term Plan for Fiber and Textile Business," *JTN*, February 1992, 19.
48. See Luckytex (Thailand) Ltd., "Luckytex," "Exporter Award, 1988," and *Luckytex Annual Report 1988–1989* (Bangkok: Luckytex, 1989).

abroad is a problem with Toyobo holding only a minority share in most of the joint ventures. Nonetheless, Toyobo imports woolen yarn from Toyobo Wool in Malaysia, and cotton fabric for sheets from Perak Textile Mills, another joint venture in Malaysia. Dusit Erawan in Bangkok provides Toyobo Japan with standard polyester/cotton fabric.[49] Toyobo itself can use much of the production of affiliates to provide materials for further finishing and marketing in Japan at Toyobo or affiliated firms, thus offsetting the loss of market share in Japan to foreign imports by importing itself from affiliates in countries with lower production costs.

Toray manages a global network of production. Toyobo has moved production offshore to offset its losses in lower-count products in markets at home and abroad. Nisshinbo has again taken a different path; it is less active abroad in spinning than any of our four target firms. Nisshinbo invested in a wholly owned subsidiary in Brazil in 1974 that today reports a workforce of six hundred and a physical base of 60,000 spindles, less than 10 percent of Nisshinbo's spindle capacity within Japan.[50] Nisshinbo has recently teamed up with the Kanematsu trading house to establish a small-scale mill in California with 23,000 spindles and one hundred air-jet looms, but Nisshinbo has been more active recently in nontextile investment abroad.[51] We can point, for instance, to merchandising and financing operations, as well as to a joint investment in plastics in Thailand and electronics in the Netherlands. One might argue that Nisshinbo lacked the capital to move production abroad in the 1960s. Extensive investment offshore in the next decade would have demanded considerable commitment of funds for this midsize spinning firm. The more recent investments in nontextile areas abroad suggest the firm may be looking to investment outside Japan to support its diversification strategy, rather than to expand its textile production.

Tsuzuki Spinning has taken its own path in investing abroad in textile production. Tsuzuki tends to establish wholly owned subsidiaries,

49. *Toyobo Annual Report 1991*, 5.

50. Toyo Keizai, *Gyōshu betsu kaigai shinshutsu kigyō 1990* (Enterprises investing abroad: Investments listed by industries) (Tokyo: Toyo Keizai Data Bank, 1991), and *Kaisha betsu kaigai shinshutsu kigyō* (Enterprises investing abroad: Investments listed by firm) (Tokyo: Toyo Keizai, 1991).

51. Nihon Sen'i Kyokai, *Sen'i nenkan 1990*, 42; "Nisshinbo California Inc. to Start Soon Full Operation of Its Spinning and Weaving Factory, the First One in the West," *JTN*, February 1990, 32–34.

rather than joint ventures abroad. Industrial Textil Tsuzuki was established in São Paulo in 1960, and today employs twelve hundred workers. But then Tsuzuki turned to the United States and established TNS Mills with headquarters in South Carolina in 1967, employing 2200 workers today, about half the size of Tsuzuki's workforce in Japan. The U.S. firm now has five mills spread across the state of South Carolina, and others in Georgia and North Carolina. Tsuzuki has recently been expanding in the European market. A textile subsidiary in Portugal operates 31,000 spindles, and an Austrian mill operates 41,500 spindles, and the firm plans to open a mill in Spain in the latter part of 1992.[52] Two features of Tsuzuki Spinning's investments diverge from the pattern of the mogul investment abroad: wholly owned subsidiaries, and location in advanced industrialized nations.

Again we might ask, what accounts for the differences among the four firms in offshore production, and what effect will the variations have on the industry-wide process of adjustment? Strategies of foreign investment reflect the goals and resources of the firm. Toray has committed itself to a global network of production and marketing in textiles, and the firm has the resources and expertise in multifiber production necessary for the task. Moreover, the expense and volatility of raw materials for synthetics production encourage the search for production closer to resources. Toray can rely on various incentives from local states in developing economies anxious to develop both chemical and textile industries. Toyobo, however, may be a "comprehensive textile maker," but it cannot command the resources or expertise comparable to Toray for foreign investment. Local firms in developing societies can develop the necessary expertise to run imported spinning machinery from Toyoda, Nissan, or elsewhere. Toyobo can offer expertise in running the machinery, but its more important contribution is in organizing the production process, helping in procurement of raw materials, and in integrating weaving, finishing, and dyeing processes. Toyobo can plays a part in production offshore at joint ventures, but with limited capital resources and with expertise largely limited to spinning and blending yarns.

Nisshinbo stands apart from the four firms here in the small scale of its investment in production abroad. Even more limited than Toyobo

52. "Tsuzukibo: Seikai senryakuga kyūsoku shinten" (Tsuzukibo: A world strategy of rapid advance), *Nihon Sen'i Shimbun*, 12 September 1989, 3; "Editorial: Japanese Spinners Invest Abroad Actively," 16.

in capital resources and textile expertise beyond spinning and finishing, Nisshinbo has decided to upgrade its technology and focus on production within Japan. Tsuzuki, however, has been among the most active of the spinners in investment abroad in recent years. From interviews with executives at the firm and others in the industry, it appears Tsuzuki is exporting its know-how, specifically its TNS production system in the production of middle-count yarns and fabrics. The firm's strategy in the United States has brought production almost to the edge of the cotton fields, reducing the cost of raw materials. Production within the United States and Europe permits quick and inexpensive access to a variety of very attractive markets. Tsuzuki Spinning has ambitions to become the world's leading spinner.

Has the transition in textiles been successful? Financial records across the decade chronicle growth in sales and incomes. Annual sales at Toray grew some 10 percent, and net profits better than doubled. Sales at Toyobo grew 28 percent, and net profits better than doubled. Sales at Nisshinbo grew 25 percent, net profits 50 percent. The annual value of sales at Tsuzuki grew nearly 50 percent over the decade, and it seems logical to assume that profits grew as well. Toyobo and Nisshinbo recorded amounts of value added and ratios of sales to value added in 1990 in line with those of the wider textile industry. But in comparison to other manufacturing sectors, the textile industry recorded only about half the average amount of value added and also fell below the average in ratio of sales to value added across all manufacturing sectors. The synthetics industry, to include Toray, also fell below the average of other industries in ratio of sales to value added, although the total amount of value added for the synthetics industry far exceeded the average for all industries. The evidence suggests survival and even profits, but little prospect for profitable expansion in textile production at home for the moguls.

What do the firms tell us of direction and divergence? Adjustment continues at the firms. We find a vision or blueprint for change crafted by a coalition of industry and state officials, encouraging product specialization, diversification, and shifting of production offshore. A review of the four firms documents differences in adjustment strategies and prospects for growth. Architects of the "vision" statements time and again called for cooperation between industry and state to moderate fluctuations in the market during the adjustment, whether in pressing for "orderly imports," or in uniform implementation of production cartels and scrapping programs to bring supply in line with demand. Collective efforts during this vulnerable period of adjustment in the production cartels and scrapping programs would ensure predictability in market share, and at least consistent, although reduced earnings. We have noted how mavericks, and to a lesser extent even traditional moguls like Nisshinbo moved away from the industry consensus by expanding capacity and market share. That most of the moguls complied with suggested directions of change gave meaning and force to the "compass" provided in the state/industry vision for the restructuring. The fact that dissenters and mavericks broke ranks and adhered only to their own market compass revealed seams in this well-woven garment of collective efforts for reforming the industry. A centripetal force among the textile mills is evident in how they acted as a community to moderate markets and gain state support for communal efforts. But the same mills also face a centrifugal force of market ties within and beyond Japan, prompting firms to act as individuals to gain market share.

More radical strategies of adjustment raise questions about the industry and change. How effectively can moguls long identified with spinning or synthetic production move into nontextile areas of investment? The plant endowment, technology, expertise, and experience of these firms remains firmly within the textile industry. Can a firm efficiently transfer such resources to other product lines? One might also ask about the efficient scale of offshore investment in textile production, given the constraints on capital investment for the highly leveraged spinning moguls and the limits on their expertise in organizing global production and marketing activities. Looking within adjustment programs to problems of consensus, we find strategies of diversification eroding the identity of the industry itself as firms compete to survive.

4 The State

I began this book by discussing how public and private interests can clash in the process of industrial adjustment. I noted corporatist strategies that bridge public and private interests and encourage coalitions between labor and capital to find benefit in more encompassing or comprehensive goals. Interest organizations must find ways to mobilize among their individual members a joint interest that serves but is not limited by the special interests of any one member. A state intent on shaping an effective consensus among the organized interests of capital and labor in any one industry must likewise work to establish more encompassing interests within and across sectors. McKean has written of encompassing organizations in Japan that extend or "push out" the time horizons of special interests in order to establish long-range perspectives and avoid wasting resources on short-term advantages.[1] Encompassing organizations among state bureaucracies and the organized interests of capital and labor are the key to successful industrial policy, rather than simply a strong state. One might add they must extend not only their time horizons, but also the breadth of interests among their members to ensure commitment to organizational goals and a basis for joint efforts with competing organizations.

1. Margaret A. McKean, "State Strength and the Public Interest," in *Political Dynamics in Contemporary Japan*, ed. Gary D. Allinson and Sone Yasunori (Ithaca: Cornell University Press, 1993), 9. Ron Dore writes more simply of the art of compromise: "The way the Japanese economy has dealt with the problem of declining industries—starting with coal and textiles in the 1960s, but on a much wider scale over the decade since the energy crisis—argues a considerable capacity for finding acceptable points of compromise between sectoral interests and the general interest." "How Fragile a Super State," in *Japan and World Depression, Then and Now*, ed. Ron Dore and Radha Sinha (London: Macmillan, 1987), 98–99.

States need firms to ensure economic security and prosperity.[2] The textile industry serves the public interest by providing multiple benefits to the national economy, satisfying local demand for yarn, fabrics, and garments. It provides employment and resources for regional development and extends the scope of affiliated local industries into export markets abroad. Such benefits prompted the Japanese state to work with the textile industry to ensure suitable market competition by establishing adjustment programs that provide an umbrella of collective funding, capacity controls, and even production controls during the years of product specialization, diversification, and offshore investment. Change in the industry did not occur without dissent within a common compass of change. Moguls and mavericks took different paths, but the divergence did not undermine a joint pattern of adjustment among the vast majority of firms in the upstream sector of the industry.

Yet how is one to explain the anomaly of dissent within a common direction? How could the state accommodate noncompliance with the industry vision among the mavericks and other dissenters and yet sustain a broader consensus among the majority of firms on directions of adjustment? The durability of the consensus suggests the state did indeed extend the length and breadth of interest perspectives among the major upstream producers, but still accommodated both the market and the market options of individual firms that opposed the textile vision. A state role of authoritative coordination offers one answer to the riddle of dissent and direction. Within a corporatist framework of tripartite negotiation, the state coordinated an adjustment process that moderated the pace of decline in the industry and fostered survival and even success among the moguls and mavericks. Coordination suggests more than simply a resolution to the antinomies of policymaker versus policy-taker. We find an erosion of state authority and resources vis-à-vis the organized interests of capital across the decades of adjustment in textiles, but we are less interested in the shifting balance of power, than in the persistence of a corporatist framework of institutionalized interests and bargaining. The state retains authority in a coordinating role as manager and monitor of the adjust-

2. Charles Lindblom, *Politics and Markets: The World's Political-Economic Systems* (New York: Basic Books, 1977); Douglass C. North, "A Framework for Analyzing the State in Economic History," *Explorations in Economic History* 16 (1979): 249–59; James O'Connor, *The Fiscal Crisis of the State* (New York: St. Martin's, 1973).

ment process, developing ideas, shaping a consensus, and working out a joint direction.

The function of "coordination" draws attention to a growing consensus between statist and corporatist theorists, which acknowledges the prominence of the Japanese state without ignoring the role of organized interests in civil society.[3] Looking to programs for "depressed industries," Merton Peck and his colleagues found the industries themselves, rather than the government, "carried most of the burdens of financing and administering the adjustment process, . . . and played an active role in planning capacity reductions." They concluded that "the government's most important role would appear to be that of coordinator and facilitator of action in the collective interest of firms in the designated depressed industries."[4]

Coordination also resonates with Samuels's thesis of "reciprocal consent," a mutual accommodation of interests between the state and organized market interests.[5] Samuels highlights the interaction of state and capital through networks of information-exchange, negotiation, and policy enforcement, but McKean turned attention from the state itself to the interplay of encompassing organizations in the tripartite

3. Gregory W. Noble, "The Japanese Industrial Policy Debate," in *Pacific Dynamics: The International Politics of Industrial Change*, ed. Stephan Haggard and Chung-in Moon (Boulder, Colo.: Westview Press, and CIS-Inha University Press, 1989), 58. The contrast between statist and corporatist emphases has been described elsewhere. For instance, Marie Anchordoguy wrote of the "Japan Inc." approach, which "emphasizes the role of domestic political structures—especially a stable and active state and a cooperative government-business relationship," versus the "Market" approach, which "holds that traditional market forces such as high savings rates, low labor costs, and heavy investment have been key factors." See her "Mastering the Market: Japanese Government Targeting of the Computer Industry," *International Organization* 42 (Summer 1988): 509.

4. Merton J. Peck, Richard C. Levin, and Goto Akira, "Picking Losers: Public Policy toward Declining Industries in Japan," *Journal of Japanese Studies* 13, no. 1 (1987): 84, 102. Michael Young also highlights the role of private enterprise in designing and implementing the restructuring process in the shipbuilding industry. See his "Structural Adjustment of Mature Industries in Japan: Legal Institutions, Industry Associations and Bargaining," in *The Promotion and Regulation of Industry in Japan*, ed. Stephen Wilks and Maurice Wright (London: Macmillan, 1991), 162–63. In a study of the declining coal industry, S. Hayden Lesbirel concludes: "MITI, in conjunction with the LDP, played an important role as manager and coordinator of the policy process." "Structural Adjustment in Japan: Terminating 'Old King Coal,' " *Asian Survey* 31 (November 1991): 1093.

5. Richard J. Samuels, *The Business of the Japanese State: Energy Markets in Comparative and Historical Perspective* (Ithaca: Cornell University Press, 1987).

formation of industrial policy. Peter Evans cited both the substantive and relational aspects of the Japanese state, both its strength and its connections with organizations of similar, comprehensive interests.[6] Corporate cohesion of the bureaucracy in Japan makes possible a state articulation of the public interest, a state expertise in macroeconomic and sectoral issues of adjustment, and state instruments for supporting, directing, and monitoring the adjustment program. For instance, state commitment to moderating the effects of unemployment and avoiding excessive competition among major firms gives content and direction to the role of coordination for a bureaucracy that must develop consensus among the disparate interests of capital and labor.

The state took part both in shaping the content of an industry vision of change and in gaining the commitment of special interests in the consensus supporting it. It was an effort that brought together state and industry on a regular basis to shape and enforce policy. The term "state" denotes both a bureaucracy and a legal order.[7] Ministries of the state bureaucracy helped design and monitor the vision. MITI has been formulating and implementing industrial policy in the textile industry since 1956 without legal authority to force compliance to the state/industry vision of adjustment. MITI's efforts coincide with what Murakami terms "indicative" state intervention: "The intervention is dependent not so much on legal penalties as on promotional measures, such as financing, tax concessions, subsidies, research and

6. Peter Evans, "The State as Problem and Solution: Predation, Embedded Autonomy, and Structural Change," in *The Politics of Economic Adjustment: International Constraints, Distributive Conflicts, and the State*, ed. Stephan Haggard and Robert R. Kaufman (Princeton: Princeton University Press, 1992), 139–81. Muramatsu Michio and Ellis S. Krauss offer a similar image of the Japanese state. "Japanese policy-making is characterized by a strong state with its own autonomous interests and an institutionalized accommodation among elites, interacting with pluralist elements." "The Conservative Policy Line and the Development of Patterned Pluralism," in *The Political Economy of Japan*, ed. Yamamura Kozo and Yasuba Yasukichi, vol. 1, *The Domestic Transformation* (Stanford: Stanford University Press, 1987), 537.

7. Michael Mann has offered a more general description of the concept of *state*, distinguishing four constituting elements: "a) a differentiated set of institutions and personnel embodying b) centrality in the sense that political relations radiate outward from a centre to cover c) a territorially-demarcated area, over which it exercises d) a monopoly of authoritative binding rule-making, backed up by a monopoly of the means of physical violence." "The Autonomous Power of the State: Its Origins, Mechanisms and Results," *European Journal of Sociology* 25, no. 2 (1984): 188.

design grants, and government contracts."[8] Diet members and their committees gave legislative authority to the visions of change. The legal order provided both a context of constraints and opportunities for structural adjustment in the textile industry and parameters for participation or "citizenship" in the industry effort.

The record of adjustment reveals an enduring pattern of negotiation between state and textile firms to moderate market fluctuations; yet an industry in decline must pay attention not only to state policies but also to market dynamics. The adjustment decades witnessed a cycle of reform and reversion, as firms complied with state plans for capacity reduction during recessions, and then reverted to full production to serve market demand in more prosperous times. The state neither protested the reversions in times of strong demand, nor did it suppress the maverick dynamic among the major firms even in periods of recession. The significance of the transition in textiles lies more in state accommodation of dissent among a somewhat dispersed sector of spinners, than in simple coordination of a consensus among a select few firms monopolizing market and production.

That is, the state's role in industrial policy can be summarized in one word, "coordination." The state negotiated a common vision based on bridging public and private interests, but to characterize this negotiation in terms such as "reciprocity" or a "symbiotic relationship" overlooks noncompliance and dissent. A porous consensus over an industry vision for change and the emergence of dissenters among the moguls and very competitive mavericks raise questions about reciprocity between state and industry in textiles. Analysis of reciprocity within a declining industry reveals conflicts between state and industry and among and within the sectors of that industry. It was through coordination of policy formation and enforcement that the state shaped an industry vision. Yet to understand this vision we must focus not solely on direction but also on divergence, not simply on consent but also on dissent, for here in these apparent antinomies lies the key to understanding the textile transition and how the state played its role of authoritative coordination.

8. Murakami Yasusuke, "The Japanese Model of Political Economy," in *The Political Economy of Japan*, ed. Yamamura Kozo and Yasuba Yasukichi, vol. 1, *The Domestic Transformation* (Stanford: Stanford University Press, 1987), 46.

Policy

How does the state shape a common or encompassing interest without denying the individual interests of firms or specific sectors of the industry? The state has sustained a common direction in textiles by securing commitment to a negotiated program of change and by accommodating dissent. There exist diverse "collective interests" in textiles, whether those of weavers or of spinners, of larger or of smaller mills, of moguls or of mavericks. The interests of the best organized and most powerful producer groups, such as the moguls, attract more attention than those of other, less well organized interests. Yet dissenters and mavericks effect their programs, not by bolting from the alliance of moguls, but rather by remaining within the leading employers' associations and carefully abiding by the letter, if not the spirit, of the procedures mandated to control capacity. The state's role in this transition is that of a recognized and formidable authority committed to stability and moderated change. The state has brought to this negotiating table its own interests in change in the industry so structured and planned to moderate effects on labor and financial institutions.

Lehmbruch has written of exclusionary negotiations between state and selected groups of capital in specific sectors, which he terms "sectoral corporatism."[9] "Sectoral corporatism" is not what was being practiced in the textile industry in Japan. The Japanese state was intent on shaping a textile policy in line with the macroeconomic priorities of the wider economy. Annual "Economic White Papers" emanating from the Economic Planning Agency (EPA) provide a context for change within a labor-intensive, light manufacturing industry such as textiles, buffeted by declining exports and growing imports.[10] Policy priorities in the 1960s included a shift toward chemical and heavy industries and trade liberalization. The firms making synthetic fibers would benefit from the emphasis on chemical industries, and the textile industry as a whole would shift toward multifiber production, but

9. Gerhard Lehmbruch, "Concertation and the Structure of Corporatist Networks," in *Order and Conflict in Contemporary Capitalism*, ed. John H. Goldthorpe (Oxford: Clarendon Press, 1984), 62.

10. JCFA, "Tsūshō seisaku bijiyon no suii" (Changes in vision evident in MITI policy), *Kassen Geppō*, August 1990, 34–35. Kosai Yutaka provides a more detailed overview of planning across the same period in his "Politics of Economic Management," in *The Political Economy of Japan*, ed. Yamamura Kozo and Yasuba Yasukichi, vol. 1, *The Domestic Transformation* (Stanford: Stanford University Press, 1987), table 2.

the rising wages in the capital-intensive chemical and heavy industries also spurred smaller wage hikes in the still labor-intensive textile industries. With the abolition of controls on imports of both textile goods and raw cotton in the 1960s, industry and state now had to respond to the effects of imported textiles on supply and demand within the domestic market.[11]

An economic plan for the 1970s promoted knowledge-intensive industries and global production. The textile industry expanded its investment in production abroad and made efforts at home to shift to more capital-intensive production of higher-count and more sophisticated blended yarns. An EPA emphasis in the 1980s on international interdependence and technological creativity prompted the mills to license foreign fashions and technology and develop specialized textile products. The plan for the 1990s linked Japan's long-term economic development to information-intensive industries, and the textile industry introduced computer-designed production processes and expanded investment in capital-intensive manufacturing. State policy extended well beyond the macroeconomic to details concerning equipment, production, and trade. Corresponding "vision" statements on adjustment in specific industries provided clear directions for restructuring.[12]

MITI and its advisory committees have sought equilibrium in supply and demand for textiles for nearly forty years. Four laws offered special help for the larger spinners to resolve problems of overcapacity. The Old Textile Law, in force from June 1956 to September 1964, provided state funding of Y1.58 billion to defray costs of scrapping some one million older spindles and prohibited the addition of new spinning capacity.[13]

11. C. E. Dillery, "Monthly Survey of Cotton, Rayon, and Synthetic Fiber Industries—Japan 1960," Agricultural Attaché, U.S. Consulate Kobe-Osaka, 14 April 1960. Record Group 166, Civil Reference Branch, National Archives.

12. Ronald Dore, *Flexible Rigidities: Industrial Policy and Structural Adjustment in the Japanese Economy, 1970–1982* (Stanford: Stanford University Press, 1991), 132–34. The coalition of state and industry continues to formulate vision statements. See "Aiming at a Life-Culture-Creative Industry: The Synthetic Fiber Industry Vision for the 21st Century," *JTN*, parts 1 and 2, March 1992, 88–92, April 1992, 130–33.

13. Sen'i Kōgyō Kōzō Kaizen Jigyō Kyōkai (The Association for the Structural Improvement Project in the Textile Industry), *Atarashii sen'i sangyō no arikata* (A new path for textile production) (Tokyo: Tsūshō Sangyōsho Seikatsu Sangyōkyoku, 1977), 44–45; Yoshioka Masayuki, *Sen'i* (Textiles) (Tokyo: Nihon Keizai Shimbunsha, 1986), 86–87; Joseph Dodson, "Quarterly Cotton Report—Japan 1963," Agricultural Attaché, U.S. Embassy Tokyo, 26 December 1963, Record Group 166, Civil Reference Branch, National Archives, 3.

The New Textile Law, in force from October 1964 to September 1970, encouraged both scrapping of old equipment and purchase of newer equipment under a replacement formula of one new spindle for two old ones. The latter law also permitted greater flexibility in the use of spindles, evidence of a trend toward a stronger industry role in adjustment, and authorized a credit fund of Y193 billion to help finance the adjustment. In sum, the law permitted insulation from market dynamics with guaranteed loans for scrapping equipment, constraints on entry of new firms, and restraints on capacity expansion.

One problem for state and mill alike was the opening of Japan's domestic market to textile imports. Despite fierce opposition from the producers, MITI insisted on the gradual opening of the market, but also fashioned legislation such as the New Textile Law to appease the producers and bridge the transition to import competition. If a leaner industry was the goal, the scrapping program failed to achieve established targets, and again the government intervened.[14] The Special Textile Law (1967–74) directed the Japan Development Bank to provide long-term, low-interest loans for equipment scrapping. Again we find an effort to foster competitiveness by reducing overcapacity in spinning of both natural and synthetic fibers, this time by scrapping 600,000 spindles.[15] The program succeeded in purchasing and disposing of 600,000 spindles by 1973.[16] Legislation included a novel call for "voluntary" disposal of an additional 400,000 spindles; here "voluntary" meant that there would be no guaranteed state loans to

14. JSA, "Annual Review of Japanese Cotton Textile Industry for the Year of 1966," *NBG*, March 1967, 2; Elden B. Erickson, "The New Textile Industry Equipment Temporary Adjustment Law," Department of State Airgram from U.S. Consul, Kobe-Osaka, 21 July 1964, 7. An evaluation of the program is cited in U.S. General Accounting Office, *Industrial Policy: Case Studies in the Japanese Experience*, Report to the Chairman, Joint Economic Committee, U.S. Congress, 20 October 1982, GAO/ID-83-11, 50–51.

15. The legislative authority for the Old Textile Law and the New Textile Law was the Law on Extraordinary Measures for Textile Industry Facilities (Sen'i Kogyo Setsubi Rinji Sochihō). Authority for the Special Textile Law was based on the Law on Extraordinary Measures for Structural Improvement in Specific Textile Industries (Tokutei Sen'i Kōgyō Kōzō Kaizen Rinji Sochihō). Sen'i Kōgyō Kōzō Kaizen Jigyō Kyōkai, *Atarashii sen'i sangyō no arikata*, 44–45; see also Brian Toyne et al., *The U.S. Textile Mill Products Industry* (Columbus: University of South Carolina Press, 1983), appendix 6D, 13.

16. Elmer W. Hallowell, "Annual Cotton Report—Japan 1973," Agricultural Attaché, U.S. Embassy Tokyo, 14 October 1973, Record Group 166, Civil Reference Branch, National Archives, 5.

cover the costs of scrapping. A trend toward greater independence of the industry and a wider industry role in capacity reduction is evident across the era of adjustment.[17]

Excess capacity continued to plague the industry, in part because the new equipment was more productive, and in part because the growing supply of cheaper, imported textiles had reduced demand. Problems in various industries such as steel, aluminum, and textiles spurred further legislation to aid restructuring. Government joined with industry once again to reduce capacity and restrict investment in additional spinning equipment. "The Depressed Industries Stabilization Law" (1978–83) was designed to encourage restructuring in declining sectors.[18] Yamazawa explains that "for each industry, a Basic Stabilization Program was established jointly by the industry, based on industry-wide consensus, and MITI to identify the extent of excess capacity by forecasting demand and supply."[19] The legislation also authorized credit to ensure necessary loans for the purpose.[20] Among its results, the program achieved the planned reductions of 15 to 18 percent in spinning capacity of nylon filament and polyester staple and polyacrylonitrile staple. An 8 percent reduction in net capacity of cotton spinning frames across the five years of the program actually exceeded the program's reduction goal of 6 percent, but still did not resolve the problems of overcapacity.

That the program failed to achieve all of its goals can be traced in part to a market recovery that resolved the very problems the program was designed to address. Market dynamics play a large part in adjust-

---

17. Paul Sheard has emphasized the central role of Japanese firms in the adjustment of the aluminum industry. The textile firms have played a similar role in the recent decades of adjustment. See his article "The Role of Firm Organization in the Adjustment of a Declining Industry in Japan: The Case of Aluminum," *Journal of the Japanese and International Economies* 5 (1991): 14–40.

18. JSA, "Annual Statistical Review of Cotton and Allied Textile Industries in Japan in 1979 and Early 1980," *NBG*, June 1980, 4. The legislation on depressed industries (Tokutei fukyō sangyō antei rinji sochihō) was in effect from May 1978 through June 1983.

19. Yamazawa Ippei, "Increasing Imports and Structural Adjustment of the Japanese Textile Industry," *Developing Economies* 18 (December 1988): 218–19.

20. Yamamura Kozo, "Success That Soured: Administrative Guidance and Cartels in Japan," in *Policy and Trade Issues of the Japanese Economy: American and Japanese Perspectives*, ed. Yamamura (Tokyo: University of Tokyo Press, 1982), 92; Robert M. Uriu, "The Declining Industries of Japan: Adjustment and Reallocation," *Journal of International Affairs* 38 (Summer 1984): 100–101; Brian Ike, "The Japanese Textile Industry: Structural Adjustment and Government Policy," *Asian Survey* 20 (May 1980): 547.

ment policies that may appear on the surface to be efforts in line with a rational and coordinated policy. Samuels concluded of the adjustment in the aluminum industry: "In sum, this case seems to be as much a desperate catch-up game of 'try' or 'die' as it is a series of rational and coordinated responses by prescient planners."[21] Indeed, the textile adjustment program did not achieve its goal of a 13 percent reduction in polyester filament spinning capacity, but recovery of demand and new products stimulated a resurgence of profitable polyester production.[22] Again we find a blend of government-supported and voluntary scrapping of capacity and further cooperation and negotiation in the difficult task of predicting demand. Government has helped both the moguls and the smaller mills to reduce capacity across the years since 1956, where cooperation between state and industry has promoted a common direction of voluntary scrapping among the larger firms, in tandem with market realities at home and abroad. Most of the major firms have complied and reduced spindle capacity, but not all.

In addition to legislation on equipment, the state also played a role in production cartels. The JSA and the JCFA gained approval from the Fair Trade Commission for formal production cartels on three separate occasions since the 1960s.[23] Clearly the Commission will be more receptive if the industry can align its petition with the established vision of adjustment published by MITI and its Textile Industry Advisory Council or, even better, gain the direct support of MITI for its petition. Initial approval and extensions permitted a production cartel on 10 percent of capacity in spinning of both synthetic and natural fibers from July 1965 until March 1967.[24] The JSA and JCFA won approval

21. Richard J. Samuels, "The Industrial Destructuring of the Japanese Aluminum Industry," *Pacific Affairs* 56 (Fall 1983), 509.

22. Peck, Levin, and Goto, "Picking Losers," 96, table 4.

23. "The Japanese Fair Trade Commission was created under the Antimonopoly Law (chap. 8) as an independent administrative commission on the model of the U.S. Fair Trade Commission. Although the commission is under the jurisdiction of the prime minister's office (Antimonopoly Law Art. 27 [2]), it functions independently of the Cabinet including MITI and the other ministries." Iyori Hiroshi, "Antitrust and Industrial Policy in Japan: Competition and Cooperation," in *Law and Trade Issues of the Japanese Economy*, ed. Gary R. Saxonhouse and Yamamura Kozo (Seattle: University of Washington Press, 1986), 65.

24. C. E. Duffy, "Restrictive Business Practices (Textiles): Termination of Spinners' Production Cartel," Department of State Airgram, U.S. Consulate Kobe-Osaka, 9 March 1967, Record Group 166, Civil Reference Branch, National Archives.

for a second cartel to reduce spinning output of filament and yarn by 40 percent between January and the end of May 1975, and the cotton spinners agreed to a voluntary production cutback of 20 percent thereafter.[25] The JSA won approval for a third cartel between April 1977 and the end of June 1978, with a goal of 25 percent reduction in production.[26] Voluntary production cutbacks among the spinners have continued through the decade of the 1980s.[27] It is important to note that the formal cartels were often followed by informal cartels, with the Federal Trade Commission approval serving as outside recognition of the severity of the problems for the industry and as a stimulus for greater cooperation among the spinners themselves in reducing production. Most but not all of the JSA members participated in the production cartels. What is particularly significant is the continuity between formal and subsequent informal cartels. The state served as the authority coordinating the reduction agreements. Mills could often maintain the restraint by themselves for a time following dissolution of the formal cartel. Indeed, the spinners now have become accustomed to voluntary constraints, which are still negotiated within the industry.

Apart from legislation and approval of formal production cartels, the state also plays a major role in the structural adjustment of the industry through trade policy. The Foreign Ministry and MITI play major roles in shaping and implementing trade policy for the industry. For instance, exporters must conform to a quota system on Japanese exports to OECD countries, and importers are subject to occasional "administrative guidance" from MITI to balance supply and demand within the domestic economy. MITI must also investigate charges of dumping raised by the local industry against importing nations. Trade issues often fall within the purview of the Foreign Ministry or MITI,

25. JSA, "Annual Statistical Review of Cotton and Allied Textile Industry in Japan, in 1976," *NBG*, January 1977, 3; Larry F. Thomasson, "Annual Cotton Report—Japan 1975," Agricultural Attaché, U.S. Embassy Tokyo, 12 September 1975, Record Group 166, Civil Reference Branch, National Archives, 5.

26. International Cotton Advisory Committee (ICAC), "Country Statements—Japan," in Secretariat, ICAC, *Country Statements of Plenary Meetings* (Washington, D.C.: ICAC, 1977), 78–79.

27. JSA, "Annual Statistical Review of Cotton and Allied Textile Industries in Japan in 1982 and Early 1983," *NBG*, June 1983, 4; W. J. Child, "Annual Cotton Report—Japan 1984," Agricultural Attaché, U.S. Embassy Tokyo, 20 September 1984, Record Group 166, Civil Reference Branch, National Archives, 6; ICAC, "Country Reports 1986—Japan," and "Country Reports 1987—Japan," in Secretariat, ICAC, *Country Statements*, 89 and 34.

whereas necessary information on the penetration of imports in the domestic market and on plans for Japanese exports must come from the industry. State and industry had to cooperate in the past in the distribution of export quotas, especially to the U.S. market, but have found themselves preoccupied more recently with import policies.

The Japanese government has maintained a free-trade policy in textiles since the 1960s, refusing to impose Multi-fiber Arrangement (MFA) constraints on importers such as Pakistan, China, and South Korea. Market conformity in this case, however, may actually be supporting the state's political objective of fostering change in the local industry. Peter J. Katzenstein observes: "If Japanese producers who moved abroad in the 1950s and 1960s have recently begun, for whatever reasons, to reexport their products to Japan, the government is standing aside while the internationally oriented, stronger segments of the industry are undermining the domestically oriented, weaker ones."[28] The policy leaves MITI in the curious position of helping to fund reduction of excess capacity, yet refusing to control the growth in imports that makes the local capacity excessive in the first place. Balancing the macroeconomic priorities evident in trade policy with the sector-specific goals of restructuring a declining industry has proved difficult. For example, MITI supports the "orderly transition" of the industry in the adjustment process, but does not formally impose MFA quotas on importers to ensure an "orderly" growth of imports. One might attribute the slow growth of imports to the effectiveness of administrative guidance to textile importers, and to what Dore terms a "natural immunity" to imports within the industry.[29]

Adjustment policies in the textile industry coincide with the guidelines for "positive adjustment assistance policy" promoted by the OECD nations.[30] The state has spurned calls for overt import restric-

28. "Japan, Switzerland of the Far East?" in *The Political Economy of Japan*, ed. Takashi Inoguchi and Daniel I. Okimoto, vol. 2, *The Changing International Context* (Stanford: Stanford University Press, 1988), 282.

29. Dore explains this immunity as "a dense web of relational contracting between firms specializing in different parts of the production process, or between manufacturers and trading companies, between trading companies and retailers—relationships which are backed not only by their foundation in trust and mutual obligation, but by all the things that trust means, quality guarantees and security of supply." *Flexible Rigidities*, 248.

30. Sekiguchi Sueo, "Japan—A Plethora of Programs," in *Pacific Basin Industries in Distress: Structural Adjustment and Trade in the Nine Industrialized Economies*, ed. Hugh T. Patrick (New York: Columbia University Press, 1991), 433–34.

tions, including calls for imposition of MFA restraints. Instead, the state has offered sector-specific subsidies for equipment reductions, authorized formal production cartels, and encouraged voluntary capacity reductions. The participation of diverse industry interests within the Textile Industry Council and efforts to align reforms in textiles with macroeconomic priorities, particularly with wider trade priorities, suggests a state commitment to broad, national interests, rather than limited, special interests within a specific sector. Coordination fosters a common vision of specialization, diversification, and offshore investment, supported by reductions of equipment and production in periods of recession.

Consensus

Efforts to bring together diverse industry interests in a common commitment to change have elicited both consensus and dissent. Managing conflict and sustaining consensus appear central to Gregory Noble's profile of a state role in industrial policy. Samuels emphasizes the state's task of interest accommodation in promoting the circle of reciprocal consent with industry, and McKean writes of the state's cultivation of more encompassing interests among its bargaining partners of capital and labor. State helps industry and labor take a long-range perspective and extend the parameters of their interests to achieve positive-sum outcomes and builds consensus across three phases of policy formation: consultation, deliberation, and formalization. The process highlights the organizational bases for consensus, as well as the procedures establishing and helping to maintain commitment to a common compass of change.

Policymaking is a triangular process involving the Diet, the relevant ministries, and the industry. Major firms in the industry maintain a web of contacts with both the Diet and the bureaucracy, particularly with members of the ruling Liberal Democratic Party (LDP). Business and government are linked both by firm-specific, informal ties and by formal ties usually mediated by industry associations. A ministry charged with developing policy brokers priorities among three distinct groups: (1) individual firms; (2) organized interests of both capital and labor within the industry; and (3) influential LDP members with special expertise and interest in the area. The latter zoku, or "political tribes," of a ministry have been described as "informal groups of influential LDP Diet members clustered around the ministerial jurisdictions."[31] Even the *zoku* enjoy only limited leverage, since the relevant ministry itself drafts the legislation and must represent the wider public interest as well as the interests of industries under its jurisdiction. As one executive explained, the actual power to develop plans and devise budgets lies within the ministries, particularly at the section chief level. Diet members can negotiate and make requests, but the section chiefs actually make the budget allocations. Nonetheless, the larger firms, "speaking through" leading *zoku* concerned with MITI and industrial restructuring, have a major voice in the early shaping of legislation. The state apparently brought little of its own to the industry vision statements apart from an insistence on continuities with macroeconomic policy. I found officials in the industry associations and executives at some firms quite frustrated with the government's input to the industry vision statements. They insisted that the directions of change suggested in the statements had long been common knowledge in the industry, and that some of the firms had been responding to the changes well before publication of a common "vision." But even though the visions amounted to post-factum explanations of successful directions of change already in place among mogul firms, few criticized state oversight of the policymaking process.

MITI plays the role of coordinator in the conventional processes of "consultation and consensus formation" (*nemawashi*). A MITI official in the Textile Bureau described an extensive effort by his office to gain the opinions of various interest groups in the industry prior to drafting an industrial vision statement for the textile industry.[32] I

31. Aoki Masahiko, "The Japanese Bureaucracy in Economic Administration: A Rational Regulator or Pluralist Agent?" in *Government Policy towards Industry in the United States and Japan*, ed. John B. Shoven (Cambridge: Cambridge University Press, 1988), 271. Aoki notes further: "A Diet member becomes recognized as a zoku member corresponding to a particular ministry by acquiring knowledge, influence, and power related to the affairs of that ministry. One gains such influence by having served as parliamentary vice-minister [*gyōsei jikan*], as chairman of a subsection of the Policy Research Council [*seichōkai*] of the LDP corresponding to the appropriate ministerial jurisdiction, and in other important roles, successively. Because of their experience, zoku members have gained considerable expertise and access to information regarding the activities and affairs of the relevant ministries."

32. Charles McMillan has offered a topology of the "portfolio approach" to sectors evident at MITI. See his book *The Japanese Industrial System* (Berlin: Walter de Gruyter, 1989), 80–83.

asked various people at MITI and in the industry about the drafting of the initial proposals. Officials at the labor federation spoke less of direct consultation and more of MITI's endless requests for information. Leaders of the employers' organizations echoed the sentiment of labor leaders and pointed to consultation in areas of special expertise, but the executives at the firms recalled a more intensive process of consultation, both between individual firms and MITI and among committees of representatives from various firms. Coordination rather than control best typifies the state's role at this stage of the process, although leaders among labor and capital did note that once a policy was drafted and submitted to the advisory council, very few changes were possible. Consultation ensures that multiple interests are heard, that various organized interests have a part in the process. Consultation is an effort to broaden and extend special interests into a more encompassing industry interest and also to engender commitment as well as consensus.

Both commitment and consensus are nurtured in the next step of deliberation. Here we find MITI fostering the long-term interest of the industry, rather than the short-term advantages of any one sector. The Textile Industry Council (Sen'i Kōgyō Shingikai) has responsibility for revising policy before submitting it to the legislative process.[33] Membership has varied somewhat over the years, with a total of forty-four representatives now serving on the council.[34] What has not changed is the pattern of representation by interest organizations rather than firms or individuals, and the preponderance of industry associations. Consistent participation of the associations has permitted continuity but not conformity, for they include contentious groups with disparate interests such as producers and consumers. All sectors of the industry are represented, from the JCFA all the way to the department store trade association. Former MITI officials from government-associated research institutes such as the long-standing chair, Inaba Shuzō, provide leadership and expertise for the council. Among this majority of representatives from employers' associations and the government, we

33. The proposals are also reviewed by a smaller group, the Textile Subcommittee of the Advisory Council on Industrial Structure (Sangyō Kōzō Shingikai Sen'ibu). Both this subcommittee and the larger Textile Industry Council will meet together to discuss the proposal prior to its publication.

34. Nihon Sen'i Kyōkai, *Sen'i nenkan 1990* (Textile yearbook 1990) (Tokyo: Nihon Sen'i Shimbunsha, 1990), 263.

also find two labor representatives in the group, one from the Zensen and another from a smaller textile union federation. One executive from a mogul firm also serves on the Council, apparently only to show interest and lend prestige, since both the JSA and JCFA can quite adequately represent the interests of the firms. The Council also includes a few outsiders from consumer groups, the media, and the universities.

Muramatsu and Krauss write of constant attempts by the Japanese state to "coordinate the keen intra- and inter-sectoral competition" and cite the example of the advisory councils as a coordinating device to "hammer out acceptable policy solutions among competing interests."[35] Shinohara Miyohei concludes that advisory councils serve as "occasions for authorizing a bureau's or section's policy decision made in advance through informal bilateral consultations." The purpose of the councils is not debate, but rather creation of "an atmosphere of consensus around the government's views, which are often presented as 'visions.' "[36] What is the content of that consensus? Komiya Ryutaro has concluded that advisory councils generally provide a forum "in which parties can adjust proposals to reflect their joint interests."[37] My findings suggest that advisory councils limit themselves to relatively minor adjustments of the original policy drafts. I noted earlier that the input in textile policy occurred in the consultation process prior to drafting of proposals by MITI and submission to the advisory council.[38] A JSA official with long experience on the Textile Industry Council agreed there was little debate, and much effort toward gaining consensus on MITI's proposal. I find the advisory council in textiles more important for broadening interests and gaining commitment to a common policy than for representation of special interests.

35. "Conservative Policy Line," 538–39.

36. Shinohara Miyohei, Yanagihara Toru, and Kwang Suk Kim, "The Japanese and Korean Experiences in Managing Development," World Bank Staff Working Papers, no. 574, Washington, D.C., 1983, 22.

37. Komiya Ryutaro, "Introduction," in *Industrial Policy of Japan*, ed. Komiya Ryutaro, Okuno Masahiro, and Suzumura Kotaro (Tokyo: Academic Press, 1988), 18.

38. The term *shingikai* has been translated variously as "deliberative council," "advisory council," "policy council," or simply "council." I will refer to the *shingikai* as an "advisory council," even though advising is not the council's main function. Unfortunately, none of the translations adequately denotes the role of consensus formation distinguishing the *shingikai*.

Corporatist theory highlights the importance of not simply the time horizons and breadth of interests among encompassing organizations, but also of the commitment necessary across state and organized interests in a common direction. Looking beyond issues of collective action to patterns of corporatism, Philippe Schmitter argues that effective industrial policy demands "co-responsible partners in governance and societal guidance." The Japanese state has neither the authority nor the desire to force compliance to a policy directive managing adjustment in turbulent markets. The consent of industry associations and labor representatives "becomes essential for policies to be adopted; their collaboration becomes essential for policies to be implemented."[39] Consent engenders commitment, but even here there is room for noncompliance. Komiya reasons that smooth implementation is possible only because the policies reflect the negotiation of vested interests.[40] Still, implementation in textiles is not all that smooth. Within a common direction there is also dissent. Within the industry there are many different markets and interests.

Unlike a sunset industry with few market opportunities, a declining industry offers opportunities for upscale, niche markets despite losses in commodity markets. Markets as well as policy can combine to promote a more competitive textile industry. Markets may well play a larger role in the adjustment of sunset industries. Policy is reduced to damage control within the industry and encouragement of transfer of resources out of the industry, or "destructuring." For instance, S. Hayden Lesbirel concludes of the adjustment in coal, "Economic markets directed policy change, and MITI and the LDP administered it."[41] Positive policy instruments play a role, in tandem with markets, in the restructuring of a declining industry toward specialization, diversification, and offshore production. But effective policy must recognize the diverse scope of the industry interests, apparent in the affiliations of the forty-four members of the Textile Advisory Council, and how this diversity compounds the problem of negotiating vested interests. Difficulties in implementation of textile policy can be traced back to

39. Philippe C. Schmitter, "Interest Intermediation and Regime Governability in Contemporary Western Europe and North America," in *Organizing Interests in Western Europe: Pluralism, Corporatism, and the Transformation of Politics*, ed. Suzanne Berger (Cambridge: Cambridge University Press, 1981), 295.
40. Komiya, "Introduction," 18.
41. "Structural Adjustment in Japan," 1093.

the disparity of interests between smaller and larger producers, and between upstream and downstream producers, in a highly segmented industry. But there is divergence in adjustment even within a sector such as spinning.

Formalization of policy in law permits legal sanction and stipulates financial support to maintain the consensus, but laws do not provide the state with a mandate for imposing a common compass of change. Policies drafted by MITI in the advisory councils serve as the basis for legislative proposals that MITI then submits to the Diet. Muramatsu and Krauss have highlighted the growing importance of the Diet in shaping industrial policy.[42] Diet members with close ties to the textile industry, whether to firms or to labor unions, play a prominent role in committees concerned with industrial policy and other legislation pertinent to the industry. Industry sources in textiles, however, suggest that textile policy is still largely shaped by MITI and the advisory councils, for once legislative proposals have been submitted, extensive revision is just not possible. An executive from a mogul firm recalled how much pressure he had to bring to bear at the Diet to gain even a small change in a legislative proposal. Firms appear to exercise their influence in the Diet through input on a wider variety of tax, finance, and trade issues both directly and indirectly affecting the industry, but the Diet does not serve as a forum for debate on legislative proposals from MITI already reviewed by the Textile Advisory Council.

The state promotes "encompassing interests" in part through the drafting of policy, in part through the process of gaining advice, and in part through the actual commitment of organized interests to a common compass of change. If the review of policy content or "direction" suggested a prominent state role, the review of consensus formation suggests a similarly prominent but more difficult state role. Architects of change have attempted to harness market dynamics in a large and very diverse industry. The complexity of the task and the demand for specialized expertise and information in planning change have helped define the respective roles yet mutual interdependence of state and private sector in designing the adjustment process. Two scholars of adjustment concluded recently that the "character of industrial policy became passive, indicative, and intermediary for the

42. "Conservative Policy Line."

most part from 1973, rather than active, interventionist, and regulatory."[43] An earlier era of dramatic interventions and regulatory efforts may have passed, but the state is certainly not passive in the textile industry. On the contrary, we find the state an active partner in the restructuring process, representing distinctive state interests in moderating decline, rather than simply mediating disputes. A variety of policies evident in restructuring legislation, cartel authorization, and trade suggest the breadth and complexity of the state's role of authoritative coordination. The state must shape an inclusive consensus that permits constructive directions of change across an industry of disparate interests.

Dissent

Dissent has effectively undone the common compass of change in some declining industries.[44] This has not been the case in textiles. Divergence by both mogul dissidents and mavericks such as Tsuzuki Spinning did not undermine the common compass of change in textiles. Corporatist strategies of change, supported by the organized interests of capital, labor, and the state, moderated decline and fostered survival of a hierarchy of leading spinners and synthetics producers. Analysis of the roles of state, capital, and labor in shaping and enforcing common directions of change provides part of the answer, but beyond directions, we must consider how state and capital accommodate dissent without diluting their commitment to collective action.

One prominent feature of the programs was cooperation rather than direct state intervention, coinciding with Murakami's emphasis on "indicative" efforts. The state had no legal authority to force capacity reductions in any mill, but it did have considerable leeway to provide incentives to induce cooperation. MITI records indicate some 6.2 billion yen was ultimately provided in financing under the New Textile Law and 189 billion yen under the Special Textile Law, with the lion's share of state aid directed to medium and smaller enter-

43. Suzumura Kotaro and Okuno-Fijiwara Masahiro, "Industrial Policy in Japan: Overview and Evaluation," Pacific Economic Paper No. 146, Australia-Japan Research Centre, Australian National University (Canberra), 1987, 19.

44. McKean cites energy, machine tools, steel minimills, consumer electronics, and computers. "State Strength and the Public Interest," 84.

prises.[45] Larger firms could gain tax incentives and government guarantees on bank loans needed to reduce or upgrade capacity. It was clearly a cooperative endeavor in which the state played less the role of financier and more that of coordinator and promoter, since the industry needed an outside force to bring the firms to cooperate. Mark C. Tilton offers one explanation of MITI's role in shaping consensus on programs for the cement industry: "MITI helped give the trade association the discipline to make the cuts that the members wanted but couldn't coordinate on their own because of the difficulty of getting individual firms to cooperate for the sake of collective goods."[46] Once the state intervened, the firms lost some autonomy in adding capacity, but they also gained MITI's support in organizing production cartels and MITI's assistance in persuading Japanese importers to maintain orderly growth in the market.

Even state prestige and authority were not enough to ensure compliance with suggested directions of change. Mills hoping to expand during the periods of strict controls of capacity would simply purchase the registration papers for scrapped spindles from smaller firms. A market in registration papers developed as firms hoping to expand production and market share would submit purchased papers on scrapped capacity elsewhere to gain permission for adding new spindles. Production cartels offer a further example of noncompliance. A number of the firms balked at voluntary production cartels, which led the JSA to take the initiative in gaining state recognition of industry-wide cartels, and even the JSA's formal application for recognition did not include all the spinners. One could also point to noncompliance of importers who defied MITI guidance and found ways to bring in cheaper yarns and fabrics, despite the protests of the JSA. State authority fostered consensus that would not have been possible otherwise, but such authority could not mandate consensus, and seldom achieved uniform agreement on programs for capacity or production controls.

Effective implementation of the legislation and production cartels demanded more than simple compliance among the firms with plans to reduce capacity or production. Cooperation between state bureau-

45. MITI data cited in Yoshioka, *Sen'i*, 90.
46. "Trade Associations in Japan's Declining Industries: Informal Policy-Making and State Strategic Goals" (Ph.D. diss., University of California at Berkeley, 1990), 114.

cracy and industry in the program was necessary to ensure correct surveys of the scale and productive capacity of existing spindles and to formulate reliable forecasts of future market trends. Moguls and larger mavericks could provide reliable information on capacity; and given their prominence in the industry and the scale of their manufacturing base within large mills, it would have been difficult in any case to hide or disguise capacity. This was not the case for the smaller mills spread across city and countryside. Forecasting market trends was an equally daunting task. Reduction programs challenged committees of state and industry officials to predict demand across six-month periods and establish goals for scrapping capacity in view of the projected decline. But firms must also respond to markets in order to survive, and restructuring programs were not designed to insulate the industry from market demand. Yamazawa concludes that "forecasts of supply and demand, the main outcome of the Basic Stabilization Program, often differed from actual figures during worldwide fluctuations. As a result, individual firms modified their capacity decreases according to the actual figures; thus, capacity reduction was inefficient."[47] The evidence here suggests that state intervention in textiles has been active, consensual, and their forecasts at least market sensitive, if not always accurate.[48] A study of restructuring in textiles brings to light a cyclical process of cooperation in capacity reduction or production cartels in soft markets, and noncompliance in periods of stronger demand.

Dissent in textiles could also be found in the area of trade policy. The conflict of interests among various sectors over trade policy precludes direct MITI intervention on behalf of limited, "special interests" and indeed relieves MITI of the burden of taking unilateral action for the spinners versus the weavers, or vice versa. MITI must appear evenhanded in dealing with sectors within a contentious industry, for textile imports involve a wide spectrum of the textile industry, which precludes a cohesive, industry-wide negotiating position on most issues. Weavers, for instance, often prefer the lower-priced imported yarns; spinners, of course, oppose them. The state maintained its trade policy despite dissent among sectors. A common direc-

47. Ippei Yamazawa, "Trade Conflicts and Structural Adjustment," in *Economic Development and International Trade—The Japanese Model* (Honolulu: East-West Center, Resource Systems Institute, 1990), 219.

48. Daniel J. Okimoto, *Between MITI and the Market* (Stanford: Stanford University Press, 1989), 33; Katzenstein, "Japan, Switzerland of the Far East?"

tion of trade liberalization was supported by reform programs in various sectors to upgrade production to improve competitiveness.

Forging and maintaining a consistent bargaining position on trade issues is sometimes difficult even within the subsector of spinning. Spinners tend to complain about imports from Pakistan or China, where they have little investment in local mills, but not about imports from ASEAN countries, where the mills are often Japanese affiliates. If petitions for import controls have won only occasional compromises, the JSA has assumed a more direct and often successful role in negotiating market relations with major Asian trading partners. MITI is fully aware of regular negotiations between the JSA and its counterpart in Korea, that is, the Spinners and Weavers' Association of Korea, to ensure order in the import market. Similar negotiations are held with counterpart industry associations in Pakistan and China. MITI maintains its formal policy of free trade in textiles by avoiding any role in these unofficial negotiations. Yet the Japanese spinners can exercise a kind of semiofficial leverage in the negotiations by threatening to bring dumping charges to MITI for investigation and action.[49] It seems that the state may at times bridge a macroeconomic priority of liberalized trade with sector-specific interests in orderly markets by inaction, by leaving industry to press its own initiatives within a legal framework that MITI helped design.

The data suggest the state found ways to accommodate dissent on equipment controls, production cartels, and trade policy without destroying a common direction of reform articulated in the industry

---

49. The JSA filed dumping charges against both the Pakistani and Korean importers in 1983. Richard H. Friman, *Patchwork Protectionism: Textile Trade Policy in the United States, Japan, and West Germany* (Ithaca: Cornell University Press, 1990), 134–35; W. L. Davis, "Quarterly Cotton Report—Japan 1983," Agricultural Attaché, U.S. Embassy Tokyo, 11 March 1983, Record Group 166, Civil Reference Branch, National Archives, 5. Davis reported the Japanese Weavers' Association "threatened 'furious opposition' to any yarn import controls."

The Japanese knitwear manufacturers have more recently used the same strategy as the JSA to prevent dumping of Korean sweaters on the Japanese market. The Korean government formally announced a voluntary restriction of knitwear exports to Japan in February 1989, which would keep the annual growth rate under one percent for a three-year period. The Japanese Textile Federation announced on the same day that they would withdraw their request to MITI to seek countervailing duties or other antidumping measures against the Korean knitwear exports. MITI officials privately expressed relief, and publicly acknowledged the Korean decision as proof of "Korea's firm determination to maintain disciplined exports to Japan." "Editorial: Anti-Dumping Movements May Spread among Japanese Industries," *JTN*, March 1989, 15.

vision statements. What does this tell us of the state role in fostering encompassing interests within the industry? Tatsuoka Tsuneyoshi recently contrasted the role of state administrations in industrial crises in the United States and Japan. The Japanese administration takes the role of "arbitrator" who pays "attention to special interests and planned conciliatory measures before the conflict [reaches] uncontrollable proportions." Arbitration here is distinguished by close cooperation with designated interest groups such as the JSA and the JCFA. Also, arbitration is complicated by the tension between the interest of the state in stability and moderated industrial change on the one hand, and market liberalization in textiles on the other. But the state is not simply an arbitrator in this process, and certainly is not a pawn of "special interests," for the formation and maintenance of a common direction suggests the state is able to promote more encompassing, industry-wide interests among various sectors, and indeed a sector consensus among the upstream producers. Tatsuoka continues, "It (the Japanese administration) has tried to predict market forces and then create microeconomic policies aimed at smoothing and facilitating the necessary structural adjustments."[50] The record of state efforts in the restructuring of the textile industry indicates the difficulty of predicting market forces in a dispersed industry buffeted by the expansion of imports and the decline of traditional export markets.

Some would argue the state has failed to maintain enough direction to bring about effective adjustment. The seemingly endless succession of restructuring programs and the continuing need for production cartels points to the inadequacy of the restructuring effort across the wider industry, though not necessarily among the moguls and mavericks, but given the complexity of market dynamics reflecting an international division of labor, production, and consumption, and the variety of interests across the sectors of the domestic industry, the state would have needed far more authority than it has to enforce a common compass. And indeed, there was coordinated change in a common direction. Combined state/industry efforts to reduce capacity through the legislated structural adjustment programs did bring down

50. "Government Policy toward Declining Industries in Japan and the United States," Occasional Paper 89-01, Program on U.S.-Japan Relations, Center for International Affairs, Harvard University, 1989, 33.

the number of operating spindles, but not sufficiently to balance local supply with local demand. Formal and informal (i.e., "voluntary") production cartels did reduce supply and help support higher prices for yarn, but not enough to draw the industry out of its prolonged recession. Uriu concludes the government has been unable to effectively implement restructuring measures in the textile industry because it is an industry "unwilling to reallocate its resources."[51] State coordination has its limits.

The state's difficulty in maintaining a consensus sheds light on the composition of interests in this "industry" of textiles, and on the complexity of melding public and private interests. Okimoto finds that specialization by firms in a particular industry facilitated MITI's task of implementing industrial policy: "MITI officials can concentrate on developing enduring relations with a well-defined and delimited set of companies in each industry."[52] How can MITI deal with the more diversified and disparate network of firms known as the "textile industry"? The textile industry includes both producers and consumers. Spinners want a better price for their own yarn and constraints on lower-priced imported yarn, just as weavers want a competitive import market to guarantee their access to the best yarn at the lowest price. Foreign ties and product diversification have eroded the earlier identity of an industry concentrated in production of synthetic fibers and in spinning, but dispersed in weaving, dyeing, finishing, and even more so in garment making and wholesaling. Yet the moguls remain mainly textile firms and play a leading role in shaping and implementing restructuring policy.

The state has developed enduring relations with the well-defined and delimited set of moguls and their industry associations in the spinning sector of textiles. MITI has occasionally spurned the demands of the mogul spinners, particularly in trade policy, but has generally been more supportive of production cartels. That the mogul firms have largely benefited from restructuring programs reflects the state's interest in stability and moderated change in the industry. That the larger dissenters and the mavericks have not suffered from state constraints despite their defiance of the state/industry consensus on change highlights the limits of state authority, the parameters of the

51. Uriu, "Declining Industries of Japan," 106.
52. Okimoto, *Between MITI and the Market*, 126.

negotiation process, and the priority given to market dynamics as an element of both industry and state interest.[53]

A history of Japanese industrial policy in the textile transition reveals the character of Japanese corporatism. A chronicle of adjustment programs clarifies how the state has coordinated, and where it has found the strength, to moderate industrial decline. In the restructuring of the textile industry, we find the state active in efforts to manage market fluctuations. By using incentives rather than - constraints, the state assumes the role of mobilizing recalcitrant, competing firms into common action for collective interests. Coordination here suggests more than mediation or arbitration, for MITI is not simply a silent partner in the tripartite negotiation of change in the industry. Merton Peck, Richard Levin, and Goto Akira argue that high exit costs for labor accustomed to permanent employment, and for banks holding loans on the collateral of a firm's productive capacity, force the Japanese state to protect larger industrial firms from going out of business.[54] Yamazawa found evidence of such motivation in the documents of the depressed industries legislation of 1978.[55]

The state has substantive, distinctive interests in productive competition, orderly markets, and moderated decline. In textiles the state has focused more on balancing macroeconomic with microeconomic or sectoral priorities, especially evident in trade and adjustment issues, as well as on labor security and regional development. Dis-

53. "The spinning segment of the textile industry provides one example. Japanese Government officials acknowledged that it is politically difficult for MITI to recommend phasing out segments of the textile industry even though it may want to advocate this course of action." Comptroller General of the United States, *Industrial Policy: Japan's Flexible Approach*, Report to the Chairman, Joint Economic Committee, U.S. Congress, 23 June 1982, GAO/ID-82-32, 71.

54. "Picking Losers," 105. Sheard and Findlay observed the "highly articulated subcontracting networks, the intercorporate shareholdings" in Japanese corporate organization, which would likewise be disrupted by the sudden loss of a major employer and producer in an industry. See Paul Sheard and Christopher Findlay, "Japanese Corporate Organization and International Adjustment: Overview of the Issues," Pacific Economic Paper No. 165, Australia-Japan Research Centre, Australia National University (Canberra), November 1988, 8.

55. "Big and sudden changes in economic conditions, prevention of chain-repercussion of business difficulties, competition for trade policy changes, and assistance to medium and small scale firms have been listed as the justification for such intervention." Yamazawa, "Increasing Imports," 460.

tinctive, inclusive interests offer evidence of the corporate cohesion of the state. Bureaucratic organization and expertise evident at MITI offer further evidence of state strength as a knowledgeable and interested partner in the negotiation of change, but state strength alone does not explain the riddle of common direction despite dissent. What we also find is a blend of substantive and relational strengths at MITI and other government offices that permits constructive negotiations toward a consensus with the organized interests of capital and labor.

Attention to interests sheds light on the distinctive interests of the state, and on how the state bridges public and private interests. A continuum becomes apparent stretching from special, exclusive interests highlighting the advantages of individuals or single groups to comprehensive, inclusive interests characteristic of positive-sum negotiations. The focus on interests further clarifies the state role of coordination, or "conflict-management," or "consensus-formation," as an effort to extend interests to encompass more sectors within the industry, and thus promote a public interest in the survival of the industry over divisive sectoral interests. A consensus or industry "vision" is established across textiles encompassing various sectors. Within the sector of spinning, there is a "compass" of change strong enough to weather considerable divergence from the common direction of capacity reductions and production controls. Consultation and consensus formation (*nemawashi*) stretch special interests in such a way that they find benefit in more inclusive, encompassing goals.

Still, an emphasis on interests rather than on encompassing organizations might suggest the primacy of interest over organization. One might argue, for instance, that the homogeneity of Japanese cultural values fosters a commonality of interests, which in turn is reflected in the shifting and often permeable boundaries between state and civil society, one example of which could be the close, enduring ties between industry leaders and state bureaucracies evident in the Textile Industry Advisory Council, and another, the role of the JSA and the JCFA in formulating capacity reduction programs and in designing and implementing production cartels. Bernard Eccleston, for example, emphasizes the interpenetration of public and private in his definition of the state in Japan: "the central site for the negotiation of conflict over a whole range of issues, in which the state is a participant though not necessarily a consistently dominant or united one."[56] The data certainly indicate reciprocity between state and industry, and some evidence of "reciprocal consent," that is, a "mutual accommodation of state and market." Samuels concludes that such a consensus among leaders in state and industry rests on a "structure of stability," "a political stew of fragmented but stable balances within a broad conservative coalition."[57] Whether encompassing interests are cause or consequence of corporatist organization in Japan, the achievement of a consensus around more inclusive interests reinforces corporatist strategies of negotiation and organization, whereas corporatist structures make possible the formation of more inclusive interests.

A study of industrial policy also sheds light on the character of corporatist strategies and structures of adjustment in Japan. Indeed, the single theme of "embeddedness" draws together insights about state theory, collective action, and the textile adjustment.[58] The effectiveness of the consensus in textile policy can be traced not only to a strong state, or to encompassing organizations in civil society, but also to the structured negotiation of interests embedded in Japanese society. The exchange of information and expertise guiding joint efforts at predicting demand draws on actors within markets, as well as on the coordinating authority, resources, and expertise of the state. Articulation of special interests at firm and association, at enterprise union and federation, provide the substantive basis for the articulation of interest(s) common to firms within sectors and across sectors within the industry. Consensus depends not solely on either interests or organization whether in state or civil society, but rather on structured, enduring patterns of negotiation deeply rooted in society that give commitment and continuity to policy formation and enforcement.

The embedded character of corporatist strategies and structures of adjustment in textiles suggests that Japan's corporatism is "societal" because of roots in structured, long-term interests and their commitment to patterned, enduring negotiations. Inagami Takeshi argued

56. *State and Society in Post-War Japan* (Cambridge: Basil Blackwell and Polity Press, 1989), 27.

57. Samuels, *Business of the Japanese State*, 86.

58. I draw the term from Granovetter and his emphasis on the "historical and structural embeddedness" of social ties in the economy, the "specific content, history, or structural location" of economic relations. Mark Granovetter, "Economic Action and Social Structure: The Problem of Embeddedness," *American Journal of Sociology* 91 (November 1985): 486.

that the demand for enduring, patterned negotiation of interests from the organizations of capital and labor forced the state into concertation over an incomes policy and other macroeconomic efforts to defuse conflict and moderate market uncertainties with corporatist, tripartite bargaining.[59] My findings suggest a more positive role for the state in the textile adjustment, indicating bases for encompassing interests in state as well as civil society for Japan's corporatism. But if corporatist adjustment strategies and organizational structures are deeply rooted, they are not likewise equal or comparable between labor and capital, nor even across sectors of capital. Nor are such structures static in the corporatist framework of interest brokering.

59. "On Japanese-Style Neo-Corporatism: Era of a Tripartite 'Honeymoon?' " *International Journal of Japanese Sociology*, no. 1 (October 1992): 61–77.

5 The Association

Divisive dissent has crippled adjustment programs in some of Japan's depressed industries,[1] and one might conclude that a few mavericks would disrupt effective adjustment in the textile industry as well.[2] Yet corporatist strategies of adjustment in textiles moderated the pace of decline and supported survival and even success among a stable hierarchy of firms. The recalcitrant firms were "dissenters" or "mavericks," rather than "free riders" who would gain by expanding at the expense of the majority, which was cooperating in capacity reductions.[3] At the level of the industry association, there was an apparently contradictory pattern of dissent and direction. Despite clear dissent, firms within the industry associations maintained a common direction of change and somehow accommodated divergence. A review of the mediation of interests over adjustment within the industry associations helps explain how moguls and mav-

1. Gregory Noble concludes, "Consumer video and minimills are by no means the only cases that reveal a pattern of maverick firms undermining majority coalitions backed by MITI, or, especially in industries with a large number of firms, the inability to form a consensus. Within the designated depressed industries, paper linerboard and cotton spinning looked quite similar to minimills." See his article "The Japanese Industrial Policy Debate," in *Pacific Dynamics: The International Politics of Industrial Change*, ed. Stephan Haggard and Chung-in Moon (Boulder, Colo.: Westview Press and CIS-Inha University Press, 1989), 93. Richard Samuels points to "a lack of internal group discipline" as a critical factor in the poor performance of Japan's aluminum industry. See his article "The Industrial Destructuring of the Japanese Aluminum Industry," *Pacific Affairs* 56 (Fall 1983): 498.

2. Brian Ike, "The Japanese Textile Industry: Structural Adjustment and Government Policy," *Asian Survey* 20 (May 1980): 546–47; U.S. General Accounting Office, *Industrial Policy: Case Studies in the Japanese Experience*, Report to the Chairman, Joint Economic Committee, U.S. Congress, 20 October 1982, GAO/ID-83-11.

3. John McMillan, "The Free-Rider Problem: A Survey," *Economic Record*, June 1979, 95–107.

ericks established common ground and made their discrepant interests compatible.

Students of collective behavior have tried to sort out conditions promoting the commonweal over self-interest. Their model of the "prisoner's dilemma" would suggest that perpetrators of a crime interrogated separately will logically reveal what is necessary to gain the greatest benefit for themselves, rather than suffer penalties on behalf of their accomplices. Isolation of the suspects from one another, for instance, deprives them of the communication necessary for designing a common good. Fraternal continuity and discontinuity also strongly colors their decisions, for cooperation in an ongoing relationship makes possible a range of benefits not available to loners with no prospect of future contact.[4] Studies of rationality in decision-making offer insights into the policy choices of individual firms, just as analyses of communication and ongoing relationships within associations tell us of consent and dissent in an industry. Such enduring ties appear even more important in "declining" industries than in sunset or "exit" industries. Restructuring in a declining industry with possibilities for product upgrading or diversification of production prompts a far more complex and continuing dialogue than does "destructuring" in a sunset industry where all but a few firms must simply cut their losses and move on.[5]

Studies of collective rationality shed light on the structures necessary for corporatist negotiation that effectively joins individual interests with a common or public interest. Mancur Olson has argued that "encompassing organizations," with their blend of member interests yet commitment to wider societal interests, prove more effective in managing change in capitalist economies than "special interest" groups bounded by self-interest.[6] The latter impede change in the mar-

---

4. Russell Hardin, *Collective Action* (Baltimore: Johns Hopkins University Press, 1982), 3. Todd Sandler wrote of the "prisoner's dilemma" as a failure of collective action where "choices are interdependent but individual decisions are made independently." See his *Collective Action—Theory and Applications* (Ann Arbor: University of Michigan Press, 1992), 7.

5. Samuels, "Industrial Destructuring," 495–509. The author cites four sunset industries: petrochemicals, wood pulp, aluminum, and shipbuilding.

6. *The Rise and Decline of Nations: Economic Growth, Stagflation, and Social Rigidities* (New Haven: Yale University Press, 1982); "Foreword," in Sandler, *Collective Action*, vii–xvi; and *The Logic of Collective Action* (Cambridge: Harvard University Press, 1965), 127, 147.

ket by cartels and other obstructionist measures to protect themselves from competition, with little concern for sectoral interdependence or broader societal issues such as employment opportunities or regional development. A contrast in group organization and priorities between encompassing and special interests provides a tool for assessing corporatist strategies of change, particularly in studies of encompassing organizations such as "peak associations," which are critical in the tripartite negotiation over decline, where public goods such as reemployment, local real estate and industry are at stake. Given the importance of communication and continuity, we would expect the encompassing organization to represent long-term relations among partners in adjustment.

Students of both corporatism and collective behavior expect "peak associations" to include the interests of an entire industry, yet a priority on an industry "interest" does not preclude and may well encourage more encompassing concern for inter industry relations. Thus Olson argued that "encompassing organizations have some incentive to make the society in which they operate more prosperous."[7] The problem of dissent and direction in the textile adjustment, however, draws our attention not to encompassing associations alone, but to the formation of encompassing interests. A focus on peak associations in the textile adjustment would leave us solely with the Japan Textile Federation, an umbrella group including thirty-six industry associations, but it has maintained only a limited role in external relations, lobbying the Diet and bureaucracy irregularly to resolve industry crises.[8] Within the industry, it plays only a minor role in establishing consensus or dealing with dissent among the upstream producers in synthetics and natural fibers. A study of intermediation in this peak association alone will tell us little of how the big mills actually manage change.

7. *Rise and Decline of Nations*, 53. Among students of corporatism, Streeck and Schmitter have observed recently: "Corporatism requires encompassing organizations that internalize a significant part of the externalities of a group's collective action and interests, and allow for hierarchical coordination between different levels of interest aggregation and group activity." Wolfgang Streeck and Philippe C. Schmitter, "From National Corporatism to Transnational Pluralism: Organized Interests in the Single European Market," *Politics and Society* 19 (June 1991): 155.

8. The association operates with a permanent staff of only two, maintains no international affiliations, and serves solely as a lobbying unit on issues of common interest. ITMF, *The Structure of Textile Associations in ITMF Member Countries* (Zurich: ITMF, 1987), 57.

The record of associations in the textile transition suggests that special interest industry associations, rather than peak associations, help broker a common direction of change without impeding dissent and divergence among firms responding to market demand. Olson himself pointed to exceptional "special interest" organizations in Japan that represent most, if not all, members of a particular sector.[9] I examine two "special interest" associations among the mills, the JSA and the JCFA, and address the question of why they did not impede change and simply cartelize the market for their own interests. Corporatist dynamics of interest mediation among firms within a sector and across sectors gain our attention as we assess the process of melding interests among enterprise, sector, and industry. McKean concludes that encompassing organizations in Japan provide the channels for corporatist mediation of collective interests.[10] I counter that to understand this mediation one must move beyond encompassing "organizations" to a continuum between encompassing and special "interests" within industry associations, beyond channels to the programs of mediation, and beyond corporatist structures to the actual process of negotiation.[11]

Collective action to moderate or stabilize market fluctuations during periods of adjustment is the goal of restructuring programs. Cooperation that does not discourage market-oriented adjustment remains the key to effective reform. Moderating market dynamics presents a challenge to state and capital alike in joint efforts at industrial change. A porous solidarity distinguished ties among mogul and maverick firms in the JSA, where the majority of firms have maintained a common direction of adjustment despite the dissent and divergence of the minority. In contrast to divisive dissent, such flexibility at the JSA permitted a common effort of managing markets without discouraging market exposure at maverick firms, where association, state, and market all played a part in finding ways to accommodate dissent. Citizenship within the association, state constraints on collusive behavior and yet encouragement of joint efforts at adjustment, and market sensitiv-

---

9. *Rise and Decline of Nations*, 49.

10. Margaret A. McKean, "State Strength and the Public Interest," in *Political Dynamics in Contemporary Japan*, ed. Gary D. Allinson and Sone Yasunori (Ithaca: Cornell University Press, 1993), 102–3.

11. I have examined the process of dissent in greater detail in my article, "Association and Adjustment in Japan's Textile Industry," *Pacific Affairs* 66 (Summer 1993): 206–18.

ity supported accommodation of dissent without discouraging collective efforts at reform.

Industry Association

I found intense communication among members of the JSA and the
JCFA, sustained by a continuity of structure and interest linking the
two groups. A comparison of the two leading associations among
upstream producers suggests both encompass the major firms in their
respective sectors of the industry. A contrast between the two associations indicates the JSA, however, represents only a majority of spinning firms, whereas the JCFA includes all the synthetic-fiber
manufacturers. Producer associations in a declining industry provide
a fertile context for assessing dissent and direction.

The first duty of any business interest association is representation
of the collective interest of the member firms.

> Organizations (of employers and investors) do not generate power
> that does not already exist, nor do they formulate ends that do not
> derive directly from the ends that are already defined and con-
> sciously pursued at the level of the individual member firms. What
> the organization does is provide services to member firms (which in
> this way may achieve a substantial cost reduction relative to a situ-
> ation in which they would have to provide such services themselves)
> and to formulate and defend in the political area those individual
> interests (relating to taxes, tariffs, regulation of industrial relations,
> etc.) that are common to all or most member firms.[12]

The story of the moguls and mavericks offers an excellent example of
individual interests common to most but not all.

12. Claus Offe, "The Attribution of Public Status to Interest Groups: Observations
on the West German Case," in *Organizing Interests in Western Europe: Pluralism,
Corporatism, and the Transformation of Politics*, ed. Suzanne Berger (Cambridge:
Cambridge University Press, 1981), 148. For an introduction to trade associations in
Japan, see Leonard Lynn and Timothy McKeown, *Organizing Business: Trade Associ-
ations in America and Japan* (Washington, D.C.: American Enterprise Institute for
Public Policy Research, 1988); Kazunori Echigo, "Jigyōsha dantai no ruikei to kinō"
(Varieties and functions of employers' associations), *Keizai Semina*, no. 265 (February
1977): 2–10; and Atsuya Jōji, "Jigyōsha dantai no gaiyō" (An overview of employers'
associations), *Keizai Semina*, no. 265 (February 1977): 11–18.

The JSA confronted dissent among a minority of member firms regarding adjustment of spinning capacity. This was not simple disagreement, for any business association must at times pursue strategies that accommodate a minority. It was rather continuing dissent over the very direction of the industry. How did the JSA accommodate such dissent? To ensure a clear mandate, mogul executives keep a tight rein on both the JSA and the JCFA, where the top executives of the mogul firms meet regularly to approve policy and set priorities for the organizations. Division chiefs from member firms chair committees in areas of special expertise, and again themselves decide on industry-wide policies and implementation strategies. The associations promote both continuity and planned change among member firms and serve as a mirror to reflect the status of individual firms in the industry vis-à-vis their fellow firms. A comparison of the two associations highlights mogul dominance, but also nearly monopolistic control of productive capacity in the critical upstream areas of the textile industry.

The JCFA listed forty-seven member firms and sixteen associate firms in 1990.[13] Including all integrated fiber makers, the association is directed by an executive board of nineteen members and a president limited to a one-year term, which rotates among the big six synthetics firms plus Toyobo, Kanebo, and Unitika. The executive board includes firm presidents or chairs from the nine firms and a representative sample of mogul spinning firms, two smaller spinners, and a few other chemical firms with fiber-making capacity. But much of the work of the organization takes place within seven policy-area committees, twelve "special committees" focusing on day-to-day issues, and fourteen working groups active in each product area prominent among member firms. Special committees best represent the range of activities at the association and include planning, commerce, public relations, raw materials, technology, labor, research, consumer services, environmental protection, industrial hygiene, tax policy, and global environmental issues. Mogul firms on the board retain control of the JCFA and maintain links with the industry through a wider network of nonboard members and associates. A maverick like Tsuzuki Spinning is a member of the JCFA, but Kondo Spinning is not.

13. Nihon Sen'i Kyōkai, ed., *Sen'i nenkan 1990* (Textile yearbook, 1990) (Tokyo: Nihon Sen'i Shimbunsha, 1990), 271. This description of the JCFA is based on interviews with the senior executive director on 4 March 1992 and written materials about the organization received at the time.

The JSA listed a membership of sixty-two in 1990.[14] The presidency and vice-presidency of the association rotate among the mogul firms, as do the top posts on the four major working committees: the raw-cotton committee, the labor committee, the technology committee, and the government and external relations committee. Based in Osaka, the association maintains a staff of about thirty specialists to compile information and assist the committees. Committees of department chiefs from the firms draft the policies of the association, which are then deliberated and approved within an executive board of mogul presidents. Committee chairs from the mogul firms set the agenda, but non-mogul firms are represented on the committees. The JSA has recently found itself in the anomalous situation of a non-mogul firm, Tsuzuki, emerging as the largest producers of cotton yarn and fabric. Since dues at the association are based on both capacity and production, one can assume that Tsuzuki is paying the largest dues of any member, yet to date no maverick firm has been permitted to chair a committee or serve as president of the association. When I questioned this anomaly at both firms and at the association, I heard various explanations. Some suggest change is imminent. Others suggest the mavericks could not be effective in a leadership role among the moguls. And still others wonder if the maverick firms really want to be bothered with such positions.

Olson argued that associations with fewer members can mobilize for collective action more easily than larger groups, but that smaller groups with near monopolistic control of their subsector of the industry will tend to act solely on behalf of their exclusive interests.[15] The JSA and JCFA offer some contrasts in size and representation, suggesting differences of organization and interest between the subsector "industries" of spinning and synthetic-fiber making. The relative concentration, difficulty of entry or exit, and extent of horizontal and vertical integration of these industries will largely determine the structure and role of their respective industry associations. The JSA and the

14. Nihon Sen'i Kyōkai, Sen'i nenkan 1990, 283.
15. "The larger the number of individuals or firms that would benefit from a collective good, the smaller the share of the gains from action in the group interest that will accrue to the individual or firm that undertakes the action. Thus, in the absence of selective incentives, the incentive for group action diminishes as group size increases, so that large groups are less able to act in their common interest than small ones." Rise and Decline of Nations, 31.

JCFA represent the spinners and fiber producers in Japan's textile industry. The JSA includes the moguls and mavericks in a larger organization of spinners of natural fiber and multifibers representing some 80 percent of spinning capacity in the industry. Numerous smaller spinners with the remaining 20 percent of the capacity are represented separately by cooperatives within the local *sanchi* or "producing areas." But the forty-seven full members of the JCFA include all twenty-six polyester producers and twenty-one of the larger spinners, as well as all integrated fiber makers in Japan.

The difference between the comprehensive representation of all fiber makers in the JCFA, and representation of only the larger spinners in the JSA can be traced to differences in entry costs and entry controls. Huge entry costs for synthetic fiber production would have discouraged all but the largest investors even without state controls on entry. A spinner, however, can begin with a single loom and raw cotton. As for entry controls, the state has severely limited entry into synthetics and chemical fiber production, but has only occasionally, with capacity controls, discouraged entry into spinning. Different costs and controls have helped shape two industries with quite different structures. Greater concentration among the fiber makers permits the JCFA exclusive representation of the entire chemical and synthetic fiber industry, but the multiplicity of spinners leaves the JSA with representation of the majority of larger and medium-size spinners, but not of smaller groups.

One finds other differences between the two organizations. The JSA includes all firms with extensive capacity for natural-fiber spinning within Japan, including the moguls, mavericks, and also Toho Rayon. But the JSA does not include the synthetic-fiber makers or the weavers or garment makers who purchase the yarn and cloth from spinners. The JCFA, in contrast, includes both the producers and major consumers, both filament and fiber makers, and the spinners who purchase the filament and fiber for blended spinning. The two organizations in tandem monopolize associational representation of the mogul textile firms in Japan. When I asked officials at both organizations about ties between the JSA and the JCFA, they pointed to extensive communication between the JSA staff based in Osaka and the JCFA staff based in Tokyo. But a JCFA official added that the joint membership of the chief executive officers of many of the member firms on the boards of the two organizations was far more important than any staff communication. And finally we return to moguls and mavericks and our four representative firms. Toray belongs to the JCFA but not the JSA. Toyobo stands among the leaders in both organizations (Toyobo president Takizawa took his turn as president of the JCFA in 1992). Tsuzuki and Nisshinbo belong to both organizations, with Nisshinbo on the executive boards of both the JSA and JCFA, and Tsuzuki on neither.

Comparisons and contrasts between the JSA and the JCFA indicate parallels in density of communication and persistence of ties within each association, but some differences in breadth of representation. A closer look at the process of representation with labor tells us more of communication and comprehension. Labor relations remains one of the primary tasks for both associations and a touchstone for comparing how the associations represent and negotiate the interests of member firms. Committees of mogul managers from the JSA and the JCFA represent capital in industry negotiations with labor.

A collective approach to labor relations had been a priority of the JSA from its early days, but the emergence in 1947 of a national textile federation of enterprise unions in the industry gave added importance to the association's labor committee. An official of the textiles division at the Zensen was quick to point out changes over the years in the format of national bargaining between labor and capital among the major textile firms. Since 1956 capital has been represented mainly by the moguls, and usually through the mechanism of the labor committees at the JSA and the JCFA. If the committees are not directly involved in negotiations with the Zensen, the committees at least discuss the parameters of the response of individual firms to the Zensen guidelines for member unions. The Zensen publishes its own monthly report, the *Zensen*, listing the results of labor negotiations with each of the mogul and maverick firms. The margin of difference in wage raises and benefits among the moguls is remarkably small.

The JSA and JCFA also provide a forum for discussing various labor issues among the firms. A persistent shortage of labor in the industry, and the need for retraining caused by specialization within textiles or by diversification into nontextile lines has been a subject of both discussion and research at the employers' associations. Associations supply member firms with information on retraining programs and on legislation regarding retraining. Associations also maintain contacts at the Labor Ministry and MITI to keep abreast of govern-

ment programs for retraining and guidelines on relocating or terminating workers in plant closings. The associations have promoted the
shift to capital-intensive production with various reports, seminars,
and research papers about the cost and effectiveness of new technology, and here again the focus is on the retraining of workers, or the
recruitment of a more educated workforce. Associations ensure an
extensive sharing of information among the member firms, maintaining a variety and depth of communication across the sector, which
becomes a critical resource for individual firms, particularly in times
of transition.

If the study of organization suggests weaker representation at the
JSA than at the JCFA, a look at labor negotiations indicates that the
JSA does indeed comprehend a wider sectoral interest and establishes
a pattern affecting even the small firms outside the association. Comprehensive representation of capital by the JSA and the JCFA facilitates the mediation of capital's relations with labor, enabling capital to
speak with one voice in negotiation with a national federation of
labor. Consultation within labor committees at the JSA and JCFA on
guidelines for wage increments or improved benefits, or on retraining
costs make possible a common pattern of response. Within this cooperative framework, individual mogul and maverick firms can negotiate their own wage raises and try to gain a competitive advantage in
fixed labor costs over their competitors. But the margin of differences
in wage hikes recently among the moguls, and even between moguls
and Tsuzuki or Kondo Spinning has been relatively small, preventing
any one firm from gaining a major advantage in labor costs through
reduced wage and benefit packages. The pattern of assent and dissent
among member firms to restructuring policies is not evident in labor
relations. Member firms scramble to gain advantages in labor costs
through investment in labor-saving technology, rather than in dramatic differences in wages and benefits.

Sector-wide negotiations on wages and benefits is evidence of continuity and intense communication among the members of industry
associations. Apparently, the two associations encompass the interests of the two sectors. (I examine joint-consultation committees at
the firms and the industry-wide patterns of labor consultation on
adjustment programs in a later chapter.) What is clear already is that
any effort by a "free rider" to take unfair advantage of the common
accord by deep cuts in wages and benefits would draw the wrath of

both organized labor and capital. After all, citizenship has duties as well as rights, and here lies one answer to the anomaly of dissent and direction, for the dilemma of the spinners in the adjustment program is one of "citizens" rather than "prisoners." Moguls and mavericks alike are committed to the network of information-exchange, of interaction with other associations, and of lobbying with the state that the JSA provides. Exit is no more an alternative for JSA mills than is exit from one's homeland for a citizen, since there is no other home in the industry for the mills, no competing association with comparable representation.

There is more one can say about industrial citizenship. Mavericks in the JSA were loyal citizens. Firms opposing capacity reductions carefully followed established procedures by purchasing the loom registrations of contracting firms in their own efforts to expand capacity. Firms opposing production cartels still avoided dramatic production increases during the length of the cartels. Mavericks were also productive citizens and generous supporters of their association; the moguls had little ground for exiling loyal and productive citizens, despite divergence from the common compass of change. But if citizenship, state, and market defined the acceptable range and modes of dissent and consensus, they also precluded the market inefficiencies of a monopolistic association that would obstruct adaptations to changing market demands. Capacity reductions and production cartels "managed" markets to insulate firms from fluctuations in supply and demand during periods of adjustment. Moguls opted for insulation from markets; mavericks opted rather for market exposure. The JSA under mogul leadership encouraged insulation, but did not prohibit market exposure.

State

The JSA remains a "voluntary" association, rather than an imperative association a state mandates as a prerequisite for enterprise in spinning. Yet one might well argue that the state's de facto recognition of the JSA as the leading industry association for the spinners makes membership a necessity rather than a choice for individual firms. The state played a sometimes contradictory role of promoting cooperative efforts among the firms at adjustment, and yet constraining the associations from collusive behavior. For instance, the government promoted capacity reduction, but did not prohibit the purchase of loom licenses from retired equipment to be used for purchase of new equipment. They made it possible for the industry to gain exemption from antitrust laws and establish production cartels, but only with the petition of the majority of the firms, and for limited periods, and without legal sanctions on nonparticipating firms. The state helped define citizenship at the associations through constraints on, as well as incentives for, cooperation. A closer look at the conventional, state-related tasks of the associations, and at how they cooperated with the state in adjustment programs further clarify the part the state played in sustaining dissent and direction at the association.

On the government side, MITI collects information from the individual member firms on supplies, production, stock, and domestic and foreign sales. The Ministry of Finance gathers data on the financial condition of the public firms in the industry. Both the JSA and JCFA devote much of their time to government relations, serving as the industry representatives in their respective areas for state efforts in monitoring the industry. Both groups provide regular information, and respond to government requests for information on raw materials, technology, labor productivity, and foreign trade. One example of how the associations provide this information is the semiannual statistical report of the JSA, listing the spinning and weaving capacity for the mills of each of the firms.[16] It also reports semiannual production totals in yarn, mixtures, and rayon staple and includes extensive statistics on labor and trade. Monthly totals in each of the categories of the semiannual report can be found in the *Nihon Bōseki Geppō* (Monthly Report of the Japan Spinners' Association). The JCFA provides a similar service in their *Kassen Geppō* (Monthly Report of Synthetic Fibers). Much of the information found in the reports and monthly publications is gleaned from data reported directly to the state bureaucracies, not to the industry associations. It is left to the associations, however, to draw together the information of greatest interest to their members to provide a collective picture of the industry.

An exchange of information and sharing of expertise contributes to the lobbying efforts of the associations on behalf of member firms.

---

16. JSA, *Statistics on the Japanese Spinning Industry* (Bōseki jijō sankōsho) (Osaka: Japan Spinners' Association, semi-annual).

Associations serve as an institutionalized forums for cooperation between state and member firms on industry issues such as restructuring or trade. As one foreign observer noted, the industry associations "institutionalize the process" of policymaking and implementation.[17] Restructuring programs from the 1950s brought the associations into regular, intense negotiations with the Ministry of Finance, MITI, and other government offices in both policy-formation and enforcement. The Old Textile Law established a registration system for spinning equipment under government supervision. The associations played a role in (1) establishing the procedures for registration, (2) maintaining an association profile of capacity at member firms, and (3) alerting members of the progress of capacity-reduction programs. The New Textile Law and the Special Textile Law permitted a wider role for the individual firms in the scrap and build programs. Greater autonomy for the industry made the efforts of the associations even more important in providing a common picture of change for the industry. As representatives of the most important sector in the industry, i.e., producers, the associations assumed a major role in the advisory committees shaping these pieces of legislation and the state/industry visions of change.

The "Depressed Industries" legislation (1978–83) brought state and association together yet again to establish a common ground for effective capacity reduction in both natural and man-made fiber production. Here the JSA and JCFA had to initiate the process by gaining the consent of a majority of member firms, and then petition MITI to qualify for state support as a depressed industry. The two associations also provided information for the forecasts of demand and setting of reduction goals, and gathered and published information on the progress of the programs. Production cartels initiated by the JSA and the JCFA indicate a triple role for the associations: initial consensus-formation among member firms, petition for recognition by the state, and monitoring of the program. In sum, we find associations active in both policy formation and enforcement in the restructuring programs, as formerly state responsibilities in oversight of the program have devolved to the industry itself. State reliance on the associations for

17. Comptroller General, *Industrial Policy: Japan's Flexible Approach*, Report to the Chairman, Joint Economic Committee, U.S. Congress, 23 June 1982, GAO/ID-82-32, 52.

representation of the producers' interests provides recognition and legitimation for the associations among member firms and within the wider textile industry.

If the associations in tandem with the state crafted a "compass" of change requiring the reduction of capacity that won the cooperation of most of the moguls, how can we explain the opposition of mavericks and dissenting moguls? Observers reported in 1967 that moguls bought out smaller spinners so that they could then register those purchased spindles as "scrapped" and maintain their own capacity to serve growing demand in the local market.[18] Nisshinbo Industries and their affiliate, Toho Rayon, as well as Omi Kenshi, Tsuzuki Spinning, and Kondo Spinning all expanded capacity between 1967 and 1983. As for cartels, one report suggests the JSA resorted to a MITI approved cartel because of the failure of voluntary constraints promoted by the JSA itself.[19] Not all JSA members approved the market insulation strategy of production cartels, whether formal or informal. Nisshinbo heatedly objected to the production cartel in 1965, but relented in the end. An official of the JSA recalled a dreadful thirty minutes with the president of Omi Kenshi just prior to submission of the next cartel petition in the mid-1970s, when the JSA desperately needed the firm's support. The official did gain the president's reluctant assent, but not before hearing a tirade about mogul group pressure in the association. Nisshinbo refused to join the cartel in 1981, as did six other of the total JSA membership at the time of eighty-six firms.

An official history of the association documents the conflict between moguls and mavericks: "As for the spinning industry itself, in the tradition of the prewar All Japan Federation of Spinners, there was usually a clash between a majority faction seeking implementation of a curtailment, and a minority faction giving priority to the workings of the basic principle of free competition. But as in the tradition of the prewar Spinners' Association, the clash seldom surfaced in the form of direct debate, but instead served as an important dis-

18. C. E. Duffy, "Cotton Textile Industry Revival and Long-Range Reform Problems," Department of State Airgram, U.S. Consulate Kobe-Osaka, 23 February 1967, Record Group 166, Civil Reference Branch, National Archives, 3.

19. Ronald Dore, *Structural Adjustment in Japan, 1970–1982* (London: Athlone Press, 1986), 128; John M. Beshoar, "Annual Cotton Report—Japan 1981," Acting Agricultural Counselor, U.S. Embassy Tokyo, 11 September 1981, Record Group 166, Civil Reference Branch, National Archives, 4–5.

cussion topic under the moderating role of the Spinners' Association."[20] The divisions signal contrasting forces drawing the firms toward cooperation and solidarity in order temporarily to moderate market dynamics, on the one hand, and market pressures pushing the firms apart to develop their own competitive niches apart from the industry vision of change, on the other. Moguls as a group responded to a centripetal dynamic drawing firms into the JSA and its associational efforts to moderate the market during the adjustment period. Cooperative efforts fostered stability in market share during a period of adjustments for the firms within the industry, and of diversification out of the industry. Mavericks and a few dissenting moguls, on the contrary, upgraded and expanded capacity in the very same market the moguls were departing. The latter firms were responding to a centrifugal dynamic toward the market and away from cooperative efforts to moderate competition during the adjustment. Mavericks used the grace period of adjustment programs to reposition themselves for a more competitive place in an open market. State encouragement of a common compass supported the moguls, yet state constraints on collusion and monopoly discouraged sanctions against the mavericks. The state directed a tripartite process of policy formation and enforcement in which the associations were given a major role, leaving the major spinners dependent on the associations for the communication and continuity of policy and programs necessary for adjustment.

Market

State incentives for adjustment and state constraints on collusion reinforced the flexibility of citizenship within the associations, which made possible the curious blend of direction and dissent at the JSA. But state and association only contributed to a far stronger and indeed basic dynamic of market demand to which both mogul and maverick responded. Indeed, sensitivity to the market provides the rationale for state constraints and inducements, and for associational efforts at adjustment. Firms communicate and maintain continuity with one

---

20. JSA, *Bōkyō hyakunenshi* (A one-hundred-year history of the Spinners' Association) (Osaka: Nihon Bōseki Kyōkai, 1982), 143.

another only tq strengthen or maintain a market niche. Market advantage plays a part in cooperation among firms, between capital and labor, and among industries. Katzenstein has emphasized that corporatist "cooperation does not spring from altruism, but from the calculation of the long-term advantages that derive from a business community that coheres."[21] Associations represent an industry interest and work to support interdependent industries to ensure continued demand for the products their member firms produce. The final proof of communication, continuity, and encompassing interests can be found not in the association or the state, but in the market itself. Dramatic reports of threatened labor turmoil or industry "decline" periodically draw media attention to the industry associations, but their more fundamental task of maintaining productive market ties within and between industries merits little outside attention.

The maintenance of productive market ties demands a great deal of information-exchange, sharing of priorities, and bargaining. Mediation within the industry provides a common base of information for member firms on procurement on raw materials, production technology, and markets, which individual firms then use to establish and expand their competitive niche. A committee on raw materials at the JSA provides an excellent example of the cooperation and competition fostered by the association. The committee keeps contact with shippers' associations across the world, but particularly with shippers in the United States and other major supplier-nations. JSA staff members inform the committee of any changes in commercial practices or legislation affecting shipping. The Japan Cotton Traders' Association (Nihon Menka Kyōkai), with headquarters around the corner from the JSA in downtown Osaka, provides the firms and the JSA with data on raw-cotton imports based on information from member trading firms and their agents at the ports, as well as from the Customs Office.[22]

Supplied with information on supplies and trends in trade, the committee can focus on specific issues such as the problems with "sticky" cotton in bales of raw cotton from the United States. A Toyobo executive chaired the Raw Cotton Committee at the JSA in 1992 and led a

21. *Corporatism and Change: Austria, Switzerland, and the Politics of Industry* (Ithaca: Cornell University Press, 1984), 178.
22. The Japan Cotton Traders' Association publishes an annual titled *Japan Cotton Statistics and Related Data.*

JSA delegation to Washington for negotiations with the U.S. National Cotton Council over contaminated bales of high-quality U.S. cotton. Clearly, sticky cotton was a priority for Toyobo and some other leading spinners, but less so for Tsuzuki. Again I found dissent yet a common direction, and looked to the benefits for maverick and mogul in collective action. I asked the committee chair from Toyobo about the purpose and utility of such a committee. "It's like a taxi and a bus," he explained, "going alone in a taxi, or together in a bus. It is cheaper and safer in the bus." I then took the same question to an executive at Tsuzuki Spinning responsible for raw-cotton procurement who served on the Raw Cotton Committee at the JSA at the time. He admitted he seldom attended their monthly meetings because the committee focused on issues of interest only to the mogul firms. Indeed, the Tsuzuki representative had asked the committee to consider changing an agreement with the U.S. cotton shippers to permit use of ports nearer to Tsuzuki mills. The committee refused to take up the issue or adjust an earlier agreement that favored ports closer to mogul mills.

I concluded that this "bus," that is, JSA cooperation in management of raw-cotton procurement, was indeed cheaper and safer for the moguls acting in concert. Was the common bus important for the mavericks? A maverick firm like Tsuzuki needs the information provided by the committee and successful resolution of the supply problems with the United States and other countries. Reliable market and shipping information, data on sales and quantities across the industry, and accurate information on quality controls in the United States cotton industry keeps an individual firm up to date on the market and on their position vis-à-vis their competitors, since the information can be useful for decisions regarding the timing and volume of cotton purchases in the volatile raw-cotton markets. Cooperation promotes competitiveness for both moguls and mavericks. In the absence of a comparable, competing industry association, moguls and mavericks alike need the JSA. The key to solidarity in the association is not so much structure as process, not so much the institutionalized fact of compatible interests, as the ability to compromise and make interests somehow compatible.

As we turn to relations between different sectors within the textile industry, we see the limits of inclusionary efforts at the JSA and JCFA. Katzenstein has argued that the tripartite adjustment of textiles in Switzerland was made possible by the capacity of "political actors to conceive of their self-interest in broad rather than in narrow terms, and to resist the temptation of sacrificing long-term interests to short-term considerations."[23] Trade issues have also forced the Japanese producers associations to broker their sectoral interests with other interests in the industry. The JSA and JCFA occasionally send observation teams abroad, with a staff member accompanying executives of member firms, to report on trends in industries abroad.[24] Both associations regularly publish reports on industrial developments in nations considered actual or potential competitors to the Japanese firms.

The JSA and the JCFA help the firms keep abreast of their position vis-à-vis their competitors abroad and alert them to pertinent developments in trade or technology. Since the 1960s the associations have responded to abolition of controls on textile imports in Japan with extensive efforts to monitor and control the surge of foreign textiles on the local market. One problem is the diversity of producers and consumers within the domestic textile industry. A second problem is convincing counterpart associations of spinners in South Korea, Pakistan, and China to maintain an "orderly" domestic market of supply and demand in Japan. Associations must develop support within their own organizations and then within peer textile organizations before taking their petitions to MITI for controls on imports. For example, the support of the Japan Textile Federation was instrumental in gaining MITI approval in the mid-1970s to form an import council within the Textile Federation to monitor trade and report dumping.[25] Support from the Liberal Democratic Party's Textile Committee provided more extensive access to the government. Richard Friman concluded that the JSA gained state support for limited import controls only because it had built a wider alliance within the industry and with more extensive access to MITI.[26]

Melding sectoral interests into one encompassing interest across the industry presents quite a challenge. The associations won a far more

23. *Corporatism and Change*, 162.

24. See for instance JCFA, "Indoneshia o chushin toshita ASEAN shokoku no sen'i sangyō ni kansuru chōsa kenkyū" (Indonesia: A study of the textile industry in ASEAN nations) (Tokyo: Nihon Kagaku Sen'i Kyōkai, 1991).

25. Larry F. Thomasson, "Annual Cotton Report—Japan 1975," Agricultural Attaché, U.S. Embassy Tokyo, 12 September 1975, Record Group 166, Civil Reference Branch, National Archives, 5.

26. *Patchwork Protectionism: Textile Trade Policy in the United States, Japan, and West Germany* (Ithaca: Cornell University Press, 1990), 115–41.

significant concession from the government on control of imports in 1982. Parallel petitions by the weavers' association, the Zensen, the JSA, and the JCFA for protection finally gained the attention of the state. Friman reports the reaction of MITI and the Ministry of Finance to the JSA dumping suit against South Korea and against Pakistan in December 1982. Here the very consideration of the claim by the ministries gave credibility to the JSA petition and persuaded both South Korean and Pakistani exporters to agree to voluntary self-restrictions.[27] Friman attributes the spinners' success with the state in this instance to JSA efforts to mobilize a much stronger consensus across the industry for protectionist measures.[28] The effort toward "orderly imports" indicates the difficulty yet importance of consolidating an industry-wide policy on imports and the growing importance of the JSA role in negotiating trade policy in its semiannual meetings with spinners and exporters from South Korea, China, and Pakistan.

The bartering of intra- and intersectoral interests also highlights the prior task of intermediation at the associations. The review sheds light on the process of making diverse interests compatible and tells us still more of emerging patterns of dissent and direction. Cooperation and competition within the industry among spinners, between spinners and weavers, and between producers and merchants raise the issue of persisting solidarity or citizenship for the spinners. Spinners have a collective interest in assuring that their shared priorities gain a hearing in negotiations with state, labor, and the wider industry, but the spinners themselves are changing. Japan's mills spin natural and synthetic fibers into yarns, and undertake some weaving of fabric as well. The growing sophistication of textile production has resulted in a variety of cotton-based products and specialty yarns and has reduced emphasis on mass production of commodity yarn. How will this affect the JSA? Products strongly influence the interests of producers. A variety of cotton-based products suggests a greater variety of inter-

27. JSA, "Annual Statistical Review of Cotton and Allied Textile Industries in Japan in 1983 and First Half of 1984," *NBG*, August 1984, 4.

28. See *Patchwork Protectionism*, chap. 5, "Japanese Textile Trade Policy," 115–41. See also W. L. Davis, "Quarterly Cotton Report—Japan 1982," Agricultural Attaché, U.S. Embassy Tokyo, January 1982, and his "Quarterly Cotton Report—Japan 1982," Agricultural Attaché, U.S. Embassy Tokyo, 20 December 1982, Record Group 166, Civil Reference Branch, National Archives, 7. Also see Ronald Dore, *Flexible Rigidities: Industrial Policy and Structural Adjustment in the Japanese Economy, 1970–1982* (Stanford: Stanford University Press, 1991), 129.

ests among JSA members, which is quite different from the earlier unity of interest in the promotion of cotton-yarn production and sales. Yet even textiles in general may provide a sufficient focus of interests for effective action by an industry association. A recent statement by the chair of the association suggests control of imports will remain a leading priority and perhaps an effective although negative focus of interest.[29]

A more significant problem for the industry is investment abroad and diversification into nontextile products. An emphasis on reducing capacity and controlling production in cotton spinning has proved divisive among JSA members. Mogul and maverick responses indicate a conflict of interests between those still committed to local cotton spinning and those curtailing local cotton spinning. More conflicts loom on the horizon. The trend toward diversification into nontextiles does not bode well for the solidarity of a "spinning industry" in which half the sales are made in fields unrelated to cotton spinning. Investment in production abroad in various areas will likewise dilute a common interest in controlling imports, particularly as more products from Southeast Asian affiliates appear on the local Japanese market. Nisshinbo, Tsuzuki, and Kondo Spinning, for instance, do not manufacture textiles in ASEAN countries. Why should these three oppose imports from China, Pakistan, and Korea, but not from ASEAN countries where mogul firms have affiliates?

An era of even greater divergence and more powerful dissent is fast approaching, which may threaten a common compass of change. A majority of the moguls closed ranks around capacity reduction during the adjustment period. In the future these same moguls may pursue individual firm interests based on specialized products or nontextile production. Mavericks, on the contrary, with greater investment in cotton production may find themselves alone in efforts to maintain a spinning industry identity and solidarity in negotiations with the state, labor, and the wider industry. The discrepancy of interests between moguls and mavericks will challenge efforts to maintain effective patterns of cooperation and competition in Japan's fragmented, declining industries.

29. "We will press for structural changes necessary such as arrangements to realize moderate imports, including calls for the imposition of a Multi-Fiber Arrangement." "Man in Focus: Tatsuo Tanabe, Chairman, Japan Spinners' Association," *JTN*, July 1990, 72.

This chapter opened with the apparent contradiction between dissent and direction at the JSA. A majority of the member firms maintained a common direction of adjustment, including capacity reductions and production cartels, despite the noncompliance of a few firms termed "mavericks" or "dissenting moguls." Neither divisive dissent nor harmonious consensus is particularly remarkable for an adjustment program among major firms in a relatively concentrated industry. What is remarkable is the combination of dissent with persisting teamwork. Why didn't the moguls sanction or exile the mavericks, and why didn't the mavericks simply depart the association? Citizenship within the organization, a combination of state incentives for cooperation yet constraints on collusion, and access to the market may explain this curious dynamic. Citizenship, state, and market in turn help explain the formation of a sectoral priority out of the interests of individual maverick and mogul producers, and of an intersectoral interest in trade policy out of the "special interests" of diverse sectors such as spinning, weaving, dyeing, and sewing. Attention to the process of interest intermediation sheds light on how associations shape more inclusive interests out of special interests.

Mediation of interests through the industry associations within and beyond the spinning sector tell us much of corporatist strategies of adjustment. Among the few scholars of corporatism who take "complementary interests" as the problem rather than the starting point of corporatist arrangements, Harry Makler traced the formation of common interests among an industrial elite in Portugal.[30] Examination of the challenge of making discrepant interests complementary in the process of corporatist negotiation among firms within an industry association suggests citizenship within an effectively imperative association keeps both moguls and mavericks within the JSA despite opposing strategies of adjustment. A porous solidarity permits dissent without destroying a common direction of specialization, diversification, and offshore investment. Looking beyond the bargaining among firms to sectoral negotiations of capital with labor and the state, or to intersectoral negotiations across the textile industry, we find a social partnership reinforced in both structure and interest at the associa-

30. "The Portuguese Industrial Elite and Its Corporative Relations: A Study of Compartmentalization in an Authoritarian Regime," in *Contemporary Portugal: The Revolution and Its Antecedents*, ed. Lawrence S. Graham and Harry M. Makler (Austin: University of Texas Press, 1979), 125.

tion. This associational solidarity, here termed "citizenship," finds its counterpart in a societal "partnership," evident in negotiations with the Zensen or MITI over adjustment plans, or within the Japan Textile Federation over trade issues. The state played a role in promoting this inclusive approach to adjustment among capital, as did culture and tradition, and especially the market. Moguls and mavericks cooperated at the JSA to improve their market position. Spinners joined state and labor in tripartite negotiations of change to strengthen their competitive position within local and foreign markets.

Corporatism would suggest structured interests, persistent and indeed embedded through long-term ties among state, capital, and labor, with intense communication and recognition of interdependence. Olson initially raised the problem of structured business interests remaining solely "special interests," insulated from changing market dynamics by their monopoly within specific industries. We have followed Olson's emphasis on encompassing interests in industry associations as a key to effective adaptation within a corporatist framework, in line with Katzenstein's conclusion about business groups in the Swiss textile industry:

> Aware of the importance of social harmony to its overall success, the business community in Switzerland does not strictly oppose concessions to hard-pressed industrial sectors or segments of sectors. . . . [A] broad conception of self-interest reinforces the country's political stability and economic flexibility. In sharp contrast, weak business communities or weak labor movements have a narrow conception of self interest. They tend rigidly to oppose their domestic opponents and, confronted with economic threats to their existence, advocate policies inimical to adaptation, such as tariffs or job protection.[31]

Here we only begin to sort out causes in the remarkable shift from exclusion more typical of special interests, to inclusion among encompassing interests. Studies of corporatist adjustment in other industries will make possible a comparative framework for assessing the role and weight of state, market, and association in fostering inclusion.

31. *Corporatism and Change*, 163.

6   Labor's Unique Place

Workers at the mogul and maverick firms certainly have a part to play in shaping adjustment programs. They have the numbers and the resources needed too, especially in the larger firms, which have greater leverage in establishing patterns of adjustment, because it is in those very firms that unions are concentrated. About a quarter of all Japanese workers belong to unions—up to one third of all workers in industrial firms.[1] Nearly all workers at the mogul and maverick firms belong to enterprise unions, which are affiliated with the Zensen, which is in its turn a member of the Rengo coalition of federations. But despite the high unionization rates at the mills and the three organizational levels of firm, federation, and national center, and despite labor's secure place at the bargaining table with capital and the state, there is little consensus about the actual role of labor in adjustment. Yet labor rarely offers its own initiatives on the future of the industry; instead, it only responds to the initiatives of state and capital.

The weak leverage of labor in negotiations with capital has prompted some to suggest the existence of "corporatism without labor" in Japan. Yet others theorize that capital cannot afford labor turmoil in the adjustment process and thus reshapes adjustment initiatives to meet labor's priorities and maintain a cooperative atmosphere within the firm. The theme of dissent yet common direction among the moguls and mavericks finds a parallel in industrial relations, where capital must accommodate labor's dissent to maintain

---

1. Rōdō Daijin Kanbō Seisaku Chōsabu (Policy Planning and Research Department, Ministry of Labor, Japan), ed., *Rōdō kumiai kihon chōsa hokoku* (General survey on labor unions), (Tokyo: Ministry of Labor, 1992), 17, 26.

cooperation in the adjustment process. A "cooperative" model of industrial relations distinguishes corporatist strategies of industrial relations within textiles, moderating decline and fostering hierarchy among the firms. Yet the similarities in conflict resolution and interest mediation across capital and labor should not be taken as permission to ignore the differences. Industry associations bring together major textile producers of relatively equal status to broker adjustment and develop common directions of change, but industrial relations joins unequals, that is, capital and labor in efforts to plan and implement change. Discrepancies in power distinguish the tripartite structure of adjustment planning within the textile industry and Japan's model of neocorporatism beyond the industry.

Labor comes to the bargaining table without the organizational and information resources of either capital or state. Given the three features of "democratic corporatism," there is abundant evidence in Japan for an ideology of social partnership, although without a social-democratic party designing and monitoring corporatist arrangements.[2] Voluntary coordination among labor, state, and capital of conflicting objectives is not uncommon, but what one does not usually find among labor in Japan is "a relatively centralized and concentrated system of interest groups." Indeed, the cohesive interest identification necessary for coordination and centralized organization have long eluded organized labor in Japan. How then can we assess the role of labor in tripartite negotiations of adjustment in the textile industry, and specify labor's place in a model of Japanese corporatism? We might begin with organization and interest. *Organization* refers to the structures of unionized workers and to the formats of institutionalized mediation among state, capital, and labor at the levels of firm, federation, and polity.[3] *Interest* refers to the content, clarity, and cohesion of the negotiating priorities of organized labor.

Attention to organization and interest helps to distinguish two models of industrial relations pertinent to the textile industry: collusion and cooperation. The thesis of "corporatism without labor"

2. Peter J. Katzenstein, *Small States in World Markets: Industrial Policy in Europe* (Ithaca: Cornell University Press, 1985), 32.

3. I developed the theoretical basis for this argument in "Corporatism and Cooperation in Japan: Labor in Transition" (paper presented at the annual meeting of the American Sociological Association, Los Angeles, August 1994).

would coincide with the model of *collusion* where labor exercises little voice.[4] For instance, Hanami Takashi argues: "A climate of collusion [*nareai*] between the employers and the union representing the
majority of employees is the essential quality. *Nareai* is a feeling of
emotional intimacy between persons outside the kinship group. Basically the relationship is one of patronage and dependence, though the
unions frequently put on an outward show of radical militancy in
their utterances and behavior."[5] Without effective representation in
federations or national centers, and bereft of distinct, cohesive interests apart from capital and the firm, organized labor at the enterprise
opts for the survival of the firm over labor's separate priorities. Collusion would indicate the absence of any substantive voice in shaping
adjustment plans at firm, industry, or polity.

In contrast, a model of *cooperation* would suggest at least a reactive
voice and a constrained corporatist partnership for labor in the tripartite negotiation of change in textiles. The model of cooperation builds
on the "social compact" thesis of Taira Koji and Solomon B. Levine to
explain labor's newly won place in tripartite bargaining. Cooperation
is distinguished by the critical concept of "reactive voice," that is, the
capability of labor to dissent from the initiatives of capital, but not to
promote its own initiatives on reinvestment at the firm or redirection
of the industry beyond shopfloor issues of wages and working conditions. Dissent can take many forms, including posters and armbands,
leave-taking en masse, sit-downs, partial strikes, working to rule, and
slowdowns.[6] Labor conveys a more dramatic dissent through one-day,
symbolic strikes, but seldom through extended strikes. Levine has written of such leverage as "the latent power of the enterprise union to
compel management to buy peace through assuring employment security and steady advances in compensation and benefits."[7]

4. T. J. Pempel and Tsunekawa Keiichi, "Corporatism without Labor? The Japanese Anomaly," in *Trends toward Corporatist Intermediation*, ed. Philippe C. Schmitter and Gerhard Lehmbruch (Beverly Hills, Calif.: Sage, 1979), 231–70; Tsujinaka
Yutaka, "Gendai Nihon seiji no kooporatizumu-ka" (The shift to corporatism in Japanese politics), in *Kōza seijikagu* (Introduction to politics), ed. Man Uchida, vol. 3, *Seiji
katei* (The political process) (Tokyo: Sanrei Shobo, 1986), 223–62.

5. Hanami Tadashi, "Conflict and Its Resolution in Industrial Relations and Labor
Law," in *Conflict in Japan*, ed. Ellis S. Krauss, Thomas Rohlen, and Patricia G. Steinhoff (Honolulu: University of Hawaii Press, 1984), 115.

6. Douglas Moore Kenrick, *The Success of Competitive Communism in Japan*
(Houndmills, U.K.: Macmillan Press, 1988), 136.

7. Solomon B. Levine, "Japanese Industrial Relations: An External Perspective," in

The strategy of dissent or "reactive voice" owes much to the *shunto* (spring labor offensives) launched by militants to force annual wage hikes through threats of strikes and other forms of labor unrest. "Reactive voice" among moderates suggests rather dissent or disapproval within the process of policy formation and the ability to promote labor interests through renegotiation and reformulation of proposals well in advance of a final showdown. However as Taira and Levine caution: "All of this does not imply that the conservative government has now fully embraced organized labor as a full and equal partner in national economic planning, on the same basis as it seems to accept employer groups."[8] And if the state hesitates to give labor equal status in the tripartite bargaining, capital is even more reluctant to accept organized labor as an equal partner. Cohesive labor organizations at firm, federation, and national center distinguishing organized workers from the enterprise is one critical difference between cooperation and collusion. But tenuous linkages among the three levels still do not support articulation of such clear and distinct interests among workers to suggest relative equality with capital.

Reports on the results of rationalization indicate only one-fifth to one quarter of the workers were able to get early retirement, as the vast majority found work in other industries or dropped out of the workforce altogether.[9] The years of adjustment have brought dislocation to better than half of the workers in textiles. The Zensen reported a 55 percent decline in employment in the spinning industry just from 1975 to the present and a 59.percent decline in the synthetic- and chemical-fibers industry.[10] Workers in the major textile firms generally fared better than in smaller firms, because of their opportunities for transfer

---

Constructs for Understanding Japan, ed. Sugimoto Yoshio and Ross E. Mouer (New York: Kegan Paul International, 1989), 310.

8. Taira Koji and Solomon B. Levine, "Japan's Industrial Relations: A Social Compact Emerges," in *Industrial Relations in a Decade of Economic Change*, ed. Hervey Juris, Mark Thompson, and Wilbur Daniels (Madison, Wis.: Industrial Relations Association, 1985), 265.

9. Zensen, *Zensen and the Japanese Textile and Apparel Industries* (Tokyo: Zensen, 1990), 18–19; U.S. General Accounting Office, *Industrial Policy: Case Studies in the Japanese Experience*, Report to the Chairman, Joint Economic Committee, U.S. Congress, 20 October 1982, GAO/ID-83-11, 54–55.

10. Zensen, *The Seventh Joint Conference of ACTWU* [Amalgamated Clothing and Textile Workers Union], *ILGWU* [International Ladies' Garment Workers' Union], *and Zensen North Pacific Rim Textile and Clothing Unions Conference: Zensen Report* (Tokyo: Zensen, 1992), 16.

either to other mills within the company or to nontextile production lines. What role has labor played in shaping this adjustment?

The thesis of corporatism without labor suggests a compromised corporatism at best. I find rather a constrained corporatism in labor in the textile transition, wielding a protest vote on adjustment plans to ensure its interests are heard. Between labor and capital there is dissent and a common direction of change, with labor withholding consent and compliance to gain the best benefits, but relenting in the end to ensure the survival and competitiveness of the adjusting firm.

Organization

A pattern of strong enterprise unions, weak federations, and fledgling national centers among organized labor has long frustrated efforts to align the Japanese experience with the corporatism familiar to northern European nations. Some would question the logic of trying to explain labor's unique place in Japan, arguing that corporatism without labor is simply not corporatism. For instance, one scholar cites a dualist model with at least segments of labor excluded from industrial citizenship, as a counter to suggestions of a Japanese corporatism.[11] With labor divided in the dualist model, workers would exercise little voice, whereas a corporatist model implies that labor has salvaged some voice in tripartite bargaining. I have no easy resolution to the problem of labor's role in a corporatist framework. I do suggest that at least in the textile transition, there is evidence for something more than compromised corporatism in industrial relations, although admittedly less than full corporatist partnership. To answer the question of collusion versus cooperation, conformity versus reactive voice, I begin at the mills. The pattern of labor and management relations at the enterprise level imposes limits on the autonomous interest and organization of labor. The plant labor orga-

11. M. Shalev, "Class Conflict, Corporatism, and Comparison: A Japanese Enigma," in *Japanese Models of Conflict Resolution*, ed. S. N. Eisenstadt and Eyal Ben-Ari (London: Kegan Paul International, 1990), 77. See also John H. Goldthorpe, "The End of Convergence: Corporatist and Dualist Tendencies in Modern Western Societies," in *New Approaches to Economic Life*, ed. Brian Roberts, Ruth Finnegan, and Duncan Gallie (Manchester: Manchester University Press, 1985), 138–44. Goldthorpe offers only two examples of exclusion in contemporary industrialized societies: migrant labor and transient labor in small firms.

nization or "enterprise union" is the basic unit for collection of union
dues for firm, federation, and center, as well as for election of union
officials.

Membership at the enterprise union includes "all blue-collar and
white-collar workers, foremen and supervisors, and even junior management personnel."[12] Four to five full-time officials administer the
unions at the individual mogul and maverick firms, although they
maintain their employment status in the firm and return to regular
employee status upon completing their terms of office. The pattern of
bipartite negotiations in the enterprise union remains the basis for
neocorporatism in Japan's industrial relations.[13] What distinguishes
the enterprise union from the more familiar enterprise local of a trade
union is the close relationship between labor and management,
including "a high degree of both formal and informal consultation."[14]
Robert E. Cole distinguishes between consultation or deliberative
committees, and the annual negotiations over immediate issues of
wages and benefits. Negotiations focus on profit distribution, but of
greater interest here are the consultations or "Joint Consultation
Committees" (rōshi kyōgi kaigi), which take up issues of profit maxi-
mization, the growth of the firm, and adjustment.[15] The JCC serve "as
a channel for management to inform and consult with labor" on both
current conditions and future plans.[16] One might divide the process
into moments of reporting and explanation, negotiation over information and mutual shaping of proposals, and finally decision mak-

12. Kozo Kikuchi, "The Japanese Enterprise Union and Its Functions," in *Industrial
Relations in Transition: The Cases of Japan and the Federal Republic of Germany*, ed.
Tokunaga Shigeyoshi and Joachim Bergmann (Tokyo: University of Tokyo Press,
1984), 173.

13. Inagami Takeshi, "On Japanese-Style Neo-Corporatism: Era of a Tripartite
'Honeymoon?' " *International Journal of Japanese Sociology*, no. 1 (October 1992):
68.

14. John Coggins et al., eds., *Trade Unions of the World, 1989–1990* (Essex: Long-
mans, 1989), 219; Shirai Taishiro, "A Theory of Enterprise Unionism," in *Contemporary Industrial Relations in Japan*, ed. Shirai (Madison: University of Wisconsin Press,
1983), 117–41.

15. Meiyō Tetsurō, "Sen'i sangyō ni okeru kōshō sōshiki" (Patterns of negotiations
in the textile industry), *Chūrōi* (1986): 14–26. For an analysis of the roles of labor and
management, bargaining positions, and results in the 1991 negotiations see JPC, *1991
nen shuntō sōkessan* (Comprehensive settlement of the spring labor offensive for 1991)
(Tokyo: Seisansei Rōdō Jōhō Senta, 1991).

16. Robert E. Cole, *Japanese Blue Collar* (Berkeley and Los Angeles: University of
California Press, 1971), 254–55.

ing.[17] Labor and management would cooperate on each step of the
process, and then be jointly committed to the results.

The Ministry of Labor reported consultation committees active in
85 percent of manufacturing enterprises with one thousand or more
workers, with at least biweekly meetings.[18] Data on the JCCs at private firms across the nation indicate that labor exercised "reactive
voice" on issues beyond shopfloor management. For instance, topics
discussed at the committees include working conditions, personnel
matters, and management policy. Not surprisingly, labor was most
active in consultation on working conditions, including wages, retirement, safety, holidays, and work schedules, but reported little codetermination with management. Labor was less involved in
consultations on personnel matters such as hiring, transfers, and dismissals, and reported again little joint determination. But union representatives in the Joint Consultation Committees showed the least
interest in management policy, production plans, and rationalization
procedures when these were presented as only for labor's information.
The Ministry reported very little codetermination, and even little consultation on these issues.[19]

Structures and interests among the textile unions largely coincide
with the national pattern. A vast majority of textile workers at the
moguls and mavericks belong to the enterprise union of their firm,
which often includes various plant-level branches. The work of consultation committees at two leading textile firms fell in line with the
national pattern of mainly information-gathering on working conditions and some personnel matters, with less consultation and no codetermination.[20] The constitution of one such committee at a major
textile firm identified two functions: "1) to decide on various steps for

17. S. J. Park, "Labor-Management Consultation as a Japanese Type of Participation: An International Comparison," in *Industrial Relations in Transition: The Cases
of Japan and the Federal Republic of Germany*, ed. Tokunaga Shigeyoshi and Joachim
Bergmann (Tokyo: University of Tokyo Press, 1984), 162.

18. *Rōshi komyunikeision chōsa hokoku* (Report of a survey on communication
between labor and management) (Tokyo: Ministry of Labor, 1989), 43.

19. *Nihon no rōshi komyunikeision no genjō* (The realities of communication
between labor and management in Japan) (Tokyo: Ministry of Labor, 1990). The
report is summarized in Hanami Tadashi, "Worker's Participation—Influences on
Management Decision-Making by Labour in the Private Sector: Japan," *Bulletin of
Comparative Labour Relations* 23 (1992): 151–66.

20. ILO, *Social and Labour Practices of Multinational Enterprises in the Textiles,
Clothing, and Footwear Industries* (Geneva: ILO, 1984), 142, 155.

planning effective and harmonious development; and 2) to explain and discuss beforehand long-range and short-range management plans for production, sales, and stability." A subcommittee of the council was mandated "to explain and discuss beforehand the formation and changes in plans for production, investment, and personnel."[21] Management elicited labor's informed opinions on plant issues such as personnel assignments and production lines and informed union representatives on issues of firm growth, the company's financial situation, and future plans.

Labor certainly sat in on growth and adjustment planning, but had little positive input on issues beyond those concerning working conditions at the plants. Reactive voice was evident in the long, difficult negotiations over plant closings. The local unit has direct responsibility in the union hierarchy for negotiating changes in job descriptions, transfers, and wage rates. Zensen officials told me that textile unions have also come to focus much of their bargaining efforts on allied issues, including working hours, transportation, housing, and education benefits. One might conclude that the labor-management councils imply the collusion model mentioned earlier, but I would emphasize, however, that labor officials have to clarify the interests of their members and consider their bargaining positions before entering into discussions with management in the councils, as well as in preparing for wage negotiations. At the same time, the labor officials would need to learn of the ceilings on their demands, which would become evident as they learned more about the condition of the firm and industry from discussions with management. The process assumes strong bonds of trust and candor between labor and management in exchanging information and demands, but also clear interest articulation among labor leaders.

Hanami Todashi emphasizes the affective side of labor/management relations where the firm represents a kind of hierarchical kinship group characterized by bonds of emotional intimacy among all members. In contrast, Koike Kazuo concludes that "the union's interest and voice in a wide area of managerial affairs" was indeed a special characteristic of Japanese unions.[22] Hanami emphasizes the dependence of labor, just as Koike highlights its initiative. Whatever interest

---

21. Nittobo and Zensen, "Rōdō keiyaku" (Labor contract), 1 January 1992, 25–27.
22. *Understanding Industrial Relations in Modern Japan* (London: Macmillan, 1988), 252.

the union might have in the policies of the firm, I find labor's voice in the mogul and maverick firms does not extend beyond shopfloor issues within the enterprise. One can hardly deny the close ties between labor and management within the firm, but I gathered from interviews with both labor and management that although the workers do not play the adversarial role more common in Western society, they certainly do maintain distinctive interests and organizations. For instance, S. J. Park concludes that the consultation system in Japanese firms does indeed promote the participation of labor, but in a "management-centered economic system" where labor joins in the communal effort of the firm, rather than focusing on "expanding the strength of labor unions representing workers' and employees' interests."[23] Kozo Kikuchi writes similarly of a "community consciousness" among the key workers and labor union leaders, purposely reinforced by personnel policies that held out the possibility of promotion within the firm as a reward for effective, cooperative behavior.[24] Communal efforts and a corporate consciousness do not preclude distinctive labor interest and organization apart from capital, but certainly do presume a common framework in which separate interests can be negotiated.

A model of cooperation at the enterprise union provides a beginning of an answer to the problem of dissent and direction for labor within a corporatist framework. The more difficult problem for corporatist theorists is the sectoral pattern of organization among workers' unions. The relative absence of strong trade or industry federations in Japan has left scholars scrambling to explain the anomaly of corporatist organization at the enterprise and national levels, but not at the intervening level of industry federations. Yet formal federations or sectoral coalitions do exist in several key industries apart from textiles, including maritime shipping, private railways, mining, chemicals, breweries, machinery, and glassmaking, and regional groupings for taxi drivers, store clerks, and salespeople.[25] Moreover, a growing number of formal and informal coalitions have been nego-

23. "Labor-Management Consultation," 165.
24. "Japanese Enterprise Union," 193.
25. Taira and Levine, "Japan's Industrial Relations," 294; Shinoda Toru, "Tenkanki no sangyō shakai ni okeru rōdō kumiai shōshiki no tokucho—gendai Nihon ni okeru rōdō no sōshikigan nettowaku" (Distinctive features of labor union organization in a time of transition for industrial society—networks among contemporary Japanese labor organizations), *Kitakyūshū Daigaku Hōsei Ronshū* 16 (March 1989): 115–60.

tiating wages on a sectoral basis over the past two decades, parallel to the *shunto* efforts nationwide. Bargaining for adjustment strategies in declining sectors has fostered the emergence of stronger industry federations, and recently federations for coal mining unions, and for iron- and steelworkers' unions (Tekkōrōren) have gained prominence in adjustment negotiations.[26] Most interesting, of course, is the traditionally strong federation, the Zensen, and its role in adjustment of the textile industry.

It is once more necessary to distinguish between collusion and cooperation. Collusion would indicate a federation role of simply coordinating the firm-oriented interests of the enterprise. Cooperation would demand a more cohesive federation to coordinate the interests of member unions, establish clear guidelines for negotiations, and mobilize and maintain credible strategies of dissent to ensure that labor's opposition led to change in management policies. The structure and priorities of the Zensen coincide with the latter model. Employers' organizations such as the JSA and the JCFA represent the interests of capital on an industry basis. The Zensen federation of textile unions represents the interests of the enterprise unions across the industry. The JCFA maintains a monopoly of representation among the synthetics producers, and the JSA monopolizes representation for the majority of larger spinners who control some 80 percent or more of spinning capacity across the nation. The Zensen retains a monopoly of union representation at the mogul and maverick textile producers, but unlike many trade federations found in Western economies, the Zensen does not participate directly in negotiations with individual firms. Yet the federation does establish industry-wide guidelines for labor's demands, and federation leaders insist that only the Zensen has the authority to call a strike. Such authority would bring the federation into the center of the decision-making process of most labor/management negotiations, since strikes remain the ultimate weapon of a workforce for whom permanent departure from a firm is not a viable possibility.

26. Kume Ikuo, "A Tale of Twin Industries: Labor Accommodation in the Private Sector," in *Political Dynamics in Contemporary Japan*, ed. Gary D. Allison and Sone Yasunori (Ithaca: Cornell University Press, 1993), 158–80; Nitta Michio, "Conflict Resolution in the Steel Industry—Collective Bargaining and Workers' Consultation in a Steel Plant," in *Industrial Conflict Resolution in Market Economies*, ed. Hanami Tadashi (Netherlands: Kluwer Law and Taxation Publisher, 1984), 233–37.

Founded in July 1946, the Zensen Dōmei today represents 580,000 workers in 1300 unions in textile manufacturing, food and beverage sales, and discount and wholesaling operations.[27] It counts as members the vast majority of workers in the spinning and synthetic-fiber industries, with 60,800 members among workers in unions of the cotton-textile branch at Zensen, and 53,240 members among workers in unions under the chemical and synthetic-fibers branch. Like the employers' organizations, the Zensen helps give direction and definition to a textile "industry," in this case by shaping the expectations and policy of labor across the major firms. Labor union representatives from the major textile firms gather under the sponsorship of the Zensen to establish guidelines for the annual spring labor offensive on wages, working conditions, and benefits. Zensen representatives then negotiate with the labor committees of the JSA and the JCFA to establish industry-wide standards for negotiations at the firm level. Executives from the firms and officials of the employers' associations insist this is not collective bargaining, but simply discussion of the general expectations of labor. Yet labor officials add that the participants do set the guidelines for salary increases in any given year. Negotiations then move to the firms for the actual bargaining between labor and management and conclusion of labor contracts. At this point in the process, the Zensen serves as a coordinator of information, but also as an executive authority in discussions of alternatives for rejecting contracts and going out on strike.

The organization and priorities of the federation stretch beyond the local industry, for the activities of the Zensen include government relations and ties with textile labor unions in other countries. Three state issues absorb the attention of the Zensen bureaus: labor law, the annual national budget, and taxes. The state consults the Zensen much as it consults the employers' associations in gathering information and shaping policies. The Textile Bureau in the Consumers' Industries Division at MITI does not include a subsection

27. Rōdō Daijin Kanbō Seisaku Chōsabu (Policy Planning and Research Department, Ministry of Labor, Japan), ed., *Rōdō tōkei yōram 1990* (A summary of labor statistics) (Tokyo: Rōmu Daijin Kanbō Seisaku Chōsabu, 1990), 199; Zensen, "Zensen 1991," a report prepared by the International Affairs Bureau (Tokyo: Zensen, 1991), 4. The federation also has departments for mercantile unions, and local industries.

This review is based on interviews in the spring of 1992 and summer of 1993 with Zensen officials in the International Affairs Bureau, the Industrial Policy Bureau, and the Cotton Textiles' Division.

devoted to labor, but it does consult the Zensen regularly on labor issues. The Zensen has close ties with the Labor Ministry, and also works closely with the Finance Ministry and Tax Office on issues important to its membership. An exchange of information among capital, labor, and the state is critical not only for the annual wage negotiations, but also for adjustment in a declining industry.[28] Among the most important of these various responsibilities in government relations, the Zensen has been a part of the bargaining process shaping and implementing policies for the restructuring of the industry.

The federation extends the interests and organization of the local textile unions to an international network of labor confederations. Zensen joined the International Federation of Textile Workers in 1951 and hosted its World Congress in 1988.[29] It has been particularly active in common efforts with textile unions based in the United States. The Zensen joined the Amalgamated Clothing and Textile Workers Union, and the International Ladies' Garment Workers' Union in sponsoring their seventh joint conference in Tokyo in March 1992. The international focus reflects efforts of labor to gain a voice in trade issues, but also to better gauge its place in an international division of labor regarding wages, skills, and technology. Such information can then be used as necessary to strengthen the bargaining positions of labor. One finds, for instance, frequent reference in the Zensen argument for reduced working hours to "international" standards of working hours in the advanced industrialized nations, evidence of interest articulation beyond simple coordination of member unions.

Zensen efforts to influence the international division of labor are also evident in the developing nations of Asia. As the textile firms began to invest in offshore production facilities in Southeast Asia, the Zensen helped organize an umbrella organization for the textile unions in East Asia and the ASEAN countries. The federation has subsequently been a major promoter of TWARO, the Asian Pacific Regional Organization

---

28. Ron Dore points to a "trialogue about macroeconomy using statistics published by the Japan Productivity Center." *Flexible Rigidities: Industrial Policy and Structural Adjustment in the Japanese Economy, 1970–1982* (Stanford: Stanford University Press, 1991), 105.

29. "Zensen 1991," 1. The organization has since been renamed the International Textile, Garment, and Leather Workers' Federation (ITGLWF).

of the ITGLWF.[30] Whatever hopes the Zensen nurtured for a cohesive labor organization across the region, disparities in government policies toward labor, in working conditions, and in stages of economic growth have slowed the process of autonomous organization or consolidation of an independent labor voice.[31] Yet the effort to establish a regional network that could at least share information on labor conditions to offset dependence on capital appears feasible. Information exchange among national unions of varying strength is a beginning in labor's efforts to keep track of the extensive activities of Japanese textile firms in the region. Here again, the leverage of labor across national borders in the region hardly compares with the leverage of capital in ties with governments or other industrial groups abroad. Nonetheless, Zensen's organization and its commitment to establish some form of voice across borders hardly suggests collusion with capital.

Criteria of organization and interest offer evidence of cooperation and reactive voice at firm and federation. The national level, the "Japanese Trade Union Confederation" or "Rengo," established in November 1989, is now the largest labor coalition in Japan. It includes some 60 percent of all unionized workers and most of the major industrial and service industry federations. Rengo describes itself as a "national trade union center with 77 affiliates and a membership of 8 million." Rengo reported 7,615,000 members in 1992, 61 percent of the 12,397,000 unionized workers in Japan. Moreover, the concentration of Rengo members in the federations of leading industries, in manufacturing, and in larger firms gives the national center prominence in the labor movement beyond its own membership.[32] The center pro-

---

30. "Since the formation of TWARO [Textile Workers—Asian Regional Organization] in 1961, Zensen has played an active role, in terms of finance and personnel, for promoting labor movements in the textile, garment and leather industries of Asia." "Zensen 1991," 13. The formal title of TWARO is the Asian-Pacific Regional Organization of International Textile, Garment, and Leather Workers' Federations. See also ILO, *Social and Labour Practices*, 126.

31. Streeck and Schmitter comment on how difficult language barriers, inexperience, and lack of resources make it to organize national labor groups across the European Economic Community. Such problems would be compounded in Asia. See their article "From National Corporatism to Transnational Pluralism: Organized Interests in the Single European Market," *Politics and Society* 19 (June 1991): 139–40.

32. Japan Institute of Labour, ed., *Japanese Working Life Profile: Labor Statistics, 1992–1993* (Tokyo: Nissho Printing, 1992), 52; Ohara Institute for Social Research, *Nihon rōdō nenkan 1992* (Yearbook of Japanese labor, 1992) (Tokyo: Hosei University, 1992), 439.

vides a national platform for promotion of labor issues, particularly among government ministries, the Diet, and local politicians, but do the structure and priorities of Rengo provide evidence for more than collusion among enterprise firms in Japan's model of labor relations?[33]

A central committee of 126 members establishes direction for the Rengo. Five members from each of six of the major federations form the core of the committee, with the remaining representatives selected from the other seventy-three federations.[34] The JTUC Research Institute for Advancement of Living Standards, better known as the *Sōken*, serves as a think tank for the Rengo and provides much of the background research for the policy decisions. Even here the dependence of the center on larger member federations and relatively small scale of operation suggest that much of the interest articulation takes place at firm and federation. Nine researchers on loan from member federations serve as staff for the Sōken on three-year assignments.[35] The organization promotes these priorities with lobbying efforts at government ministries, through contacts with political parties in the Diet, and through its participation on the various advisory committees for national policy. Rengo has established five priorities in 1993, including economic recovery, tax reform, reform of the annual wage system,

33. It "maintains support and cooperative relations with Rengo Sangiin, a group of independents in the Upper House of the Diet, and cooperates with the Social Democratic Party of Japan, the Komeito, the Democratic Socialist Party, and the United Social Democratic Party." *Rengo*, June/July 1992, 6.

34. The eight largest are the Municipal Workers Union (Jichirō), Autoworkers Federation (Jidosha Sōren), Electrical Machinery Workers (Denkiroren), Zensen, Life Insurance Workers (Seihororen), Japan Teachers' Union (Nikkyoso), and Metal Industry Federation (Zenkinrengo), and the Telecommunications Workers (Joho-Tsushin-roren). See *Rengo*, December 1991, cover page; also Coggins, *Trade Unions of the World*, 223–34. The Central Committee convened on 20 May 1993, with the six largest member groups sending five or six delegates each, and the Metal Industry Federation, Telecommunications Workers, and the next four largest groups sending three each. One or two representatives were permitted from each of the rest of the member groups, including the regional and district offices. The invited representatives and their attendance were reported on a roster of the Rengo Central Committee, circulated on 20 May 1993.

35. Rengō Sōken (Japan Trade Union Confederation Research Institute for Advancement of Living Standards), *Rengō Sōgō Seikatsu Kaihatsu Kenkyūjo no gaiyo* (An introduction to the JTUC Institute for Advancement of Living Standards) (Tokyo: Rengo Sōken, 1993), 30. See also Rengo (Japanese Trade Union Confederation), "1993–1994 nendo seisaku, seidō yōkyū to teigen (an)" (Plan and priorities for political policy and system in 1993–1994), working papers, Central Discussion Committee (Chūyō tōron shūkai), Tokyo, 25 May 1993.

reduction of working hours, and changes in employment laws. The center usually establishes a labor position on issues already framed and debated, rather than launching their own initiatives.

Mr. Norikawa Hideaki at the Sōken cited three specific areas of Rengo lobbying efforts and their results: childcare legislation, reduction of working hours, and recently on state benefits for part-time workers. He spoke of intensive efforts since 1987 on the first two issues, but only partial results on childcare legislation, and little progress on the other two issues. Still, our immediate concern here is not with short-term successes, but with organization and clear demonstration of cohesive interests. I find the structural resources for establishing and lobbying labor's shopfloor priorities, and the clarity of their proposals indicate more than simple collusion with enterprise interests to ensure survival of the firm. A look to the process and progress of policy formation and implementation at the Rengo suggests a potentially powerful organization, momentarily hampered by radical disparities in interests across its member federations, and by limited resources. The Rengo supports the articulation and pursuit of workers' interests as identified and mobilized at firm and federation and may in time foster more wide-ranging initiatives among organized labor.

A review of three levels of interest and organization among textile workers sheds light on the substance and limits of dissent and common direction. The enterprise unions at the plant level must frame their expectations and demands within the context of a declining industry. Their central priority of job security means little without the survival and effective adjustment of the firm to stable or declining market shares. Separate labor organization and leadership helps mobilize and implement the interests of labor, both in consultations and in negotiations with management, but always in tandem with the interests of the firm and under the aegis of a firm that maintains the employment status of union leaders and holds out the possibility of promotion for employees temporarily posted to union leadership duties. Capital needs experienced, trained labor to maintain production with capital-intensive production systems, and workers need the wages, benefits, and job security of permanent employment. Enterprise unions cooperate in information-sharing and especially in adjustment strategies to maximize profits for the firm, and then compete on issues of profit distribution.

Cooperation between labor and capital on the industry level of labor federation and employee organizations parallels the patterns found within firms. Committees of enterprise union leaders from the Zensen gather with the labor committees of the JSA and the JFCA to prepare the ground for annual bargaining sessions at the firms. Federation negotiations establish a cooperative, industry-wide framework in which individual firms can compete to reduce labor costs or to upgrade the educational levels of entry-level employees by raising wages and benefits. Rapid publication of contract settlements at any one mogul firm affects bargaining positions of capital and labor at the other mogul firms, strengthening the cooperative framework of the negotiations. Zensen has secured a prominent place with capital in setting guidelines for basic, firm-level issues of wages, benefits, and working conditions, but has not won an equally prominent place in the formation of industry investment policy or the shaping of the changing identity of the industry.

The national level offers promise for strengthening the voice of labor in the polity, which would give the Zensen far more leverage with the Diet and the bureaucracy. Rengo-affiliated members of the upper house in the Diet have recently begun to exercise their influence on foreign policy issues, but the media and especially people in the textile industry have questioned whether this unwieldy coalition can find enough common ground to shape consistent policies on domestic issues.[36] Domestic policy issues usually divide producers and consumers. Zensen itself must balance the interests of textile producers among union members in the spinning and chemical-fiber industries with the interests of labor unions at the chain stores selling cheaper textile goods. The four producer divisions are often at loggerheads with the three new, and fast-growing, consumption-oriented divisions such as food, services, and distributors. I was told the consumption-oriented divisions wanted smaller wage increases in 1992 to keep the smaller firms in growing industries solvent, just as producer divisions wanted larger wage-increases from their well-established firms. If such differences impede solidarity at the smaller Zensen, one can imagine the difficulty of trying to shape policies amenable to the entire membership of the huge

36. See, for instance, Kageyama Yuri, "Rengo Challenges LDP's Grip," *Japan Times*, 16 May 1992, 3.

Rengo.[37] Yet the Rengo appears a far more promising alternative for political leverage than Zensen's identification with the minority Democratic Socialist Party. Greater leverage at the Diet and with the bureaucracy would strengthen the autonomy of labor's interests and the effectiveness of their organizations.[38]

Adjustment

A review of organization and interest offers evidence of the necessary basis for reactive voice and a pattern of cooperation in labor relations, but the question remains of how dissent and direction actually play out in labor/management negotiations over change. It is not enough simply to delineate structures and interests in the study of corporatist strategies of change. It is necessary to sort out the process of interest intermediation to complete the picture of how corporatism moderated decline, and how it fostered hierarchy and even prosperity among textile moguls and mavericks. Labor and management negotiated their separate priorities within a wider legal framework established by the state; a study of the adjustment process must begin with a brief review of state programs to support labor in a declining industry.

Lobbying by predecessors of the Rengo and other national labor groups helped persuade both the state bureaucracy and the Diet to pass legislation easing the labor transition in the textile industry. The Industry Stabilization Law of 1978 and various measures for the unemployed in specific depressed industries permitted state support for unemployment and reemployment according to both industry and region. Within depressed industries, the Ministry of Labor mandated submission and review of company plans for layoffs of one hundred or more workers. On approval of the plan, the Ministry allocated "Job-seeker Certificates" to the displaced workers, which permitted a basic living subsidy for a limited period and a

37. Shinoda Toru, "Rengō jidai ni okeru seisaku sanka no genjō to tenbō" (The present reality and future prospects of policy participation in the age of Rengō), *Nihon Rōdō Kenkyū Zasshi*, no. 379 (June 1991): 48–62.

38. Ohmi Naoto, "Rengō jidai no rōshi kankei no tenbō: neo-cooporatizumu no kannōsei?" (A view of labor/capital relations in the age of the Federation—is there a possibility of neocorporatism), *Nihon Rōdō Kyōkai Zasshi* (December 1989): 12–23.

training subsidy.[39] About 30 percent of the total number of certificates issued went to workers in the textile industry.[40] Apart from national plans, the major textile firms also gained help from local regions in their employment reduction efforts.[41] Benefits included unemployment insurance, priority for reemployment in public works, and increased efforts by local offices of the Public Employment Stabilization Bureau. The Employment Assistance Law of 1983 and a revised version in 1988 encouraged employers to retain temporarily those about to be displaced, in effect extending unemployment insurance by subsidizing wages at the firm.[42] The program also provided incentives for employers in other industries to absorb the workers displaced at the mills. Government programs helped sustain a more cooperative relationship between labor and management by establishing a safety net for the unemployed, but as Sekiguchi Sueo concludes, the ability of expanding industries such as auto manufacture to absorb the displaced steel and textile workers was more important than government programs in softening the transition.[43]

Within this context of government supports, the unions established specific priorities for labor in their declining industries. For instance, the Rengo has established three conditions for adjustments leading to dismissal of workers: "1) prior consultation and discussion with the workers; 2) the safeguarding of working conditions and the rejection of an increase in the pace of work; 3) and the guarantee of full employment and of the right to work."[44] The guidelines assume a tripartite approach to the adjustment effort, for the first two guidelines are part of labor/management negotiations within the firm or federation, and the third is a concern of the state. Zensen has also formu-

---

39. Sekiguchi Sueo, "Japan—A Plethora of Programs," in *Pacific Basin Industries in Distress: Structural Adjustment and Trade in the Nine Industrialized Economies*, ed. Hugh T. Patrick (New York: Columbia University Press, 1991), 439. See also ILO, *Social and Labour Practices*, 59.

40. C. Y. Ng, Hirano Ryokichi, and Narongchai Akransanee, eds., *Industrial Restructuring in ASEAN and Japan, An Overview* (Singapore: ISEAS, 1987), 101.

41. Shinoda Toru, "Keizai shakai henyōki ni okeru rōdō seiji—sangyō seisaku to koporatizumu" (Labor policy in a time of social and economic change—industrial policy and corporatism), *Kyūshū Daigaku Hōsei Ronshū* 15 (March 1988): 35–82.

42. ILO, *General Report of the Twelfth Session of the Textiles Committee, 1991: Report I* (Geneva: ILO, 1991), 16.

43. "Japan—A Plethora of Programs," 456.

44. ILO, *General Report*, 47.

lated more specific demands, with priorities of job security and voluntary transfers on the condition that transferred workers receive wages and benefits comparable to their earlier position.[45] In cases of dismissals, the Zensen establishes guidelines on reeducation benefits and severance benefits and closely monitors enterprise-level negotiation to maintain the guidelines. It then reports the best benefits actually obtained by unions among adjusting firms and urges unions in subsequent negotiations to get either comparable or better benefits. Enterprise unions, however, at times prove more firm-oriented than industry-oriented. Directed by lifetime employees dedicated to the prosperity of the enterprise, these unions look as much to the survival of the firm as to labor priorities. Nonetheless, the enterprise unions I visited promoted the priorities established by Zensen, and certainly the record of severance benefits carefully maintained by the Zensen indicates comparable results across mogul firms. Within this context of state programs, and union priorities, the actual process of negotiation offers further insight into reactive voice and the thesis of cooperation. I consider two separate cases of adjustment among mogul spinners: a shift to a new production line at one mill, and the closing of a second mill. Bargaining over transfers, reeducation, and terminations provide further evidence of dissent yet a common direction even in the difficult task of reassigning or reducing the labor force in a declining industry.

I was shown the text of the final agreement regarding the changeover at one mill from spinning to building materials, where labor was represented by the local branch of the spinning firm's union, with direction from the Zensen on priorities of wage and benefits for reassigned workers, and on retraining programs. Much of the document was devoted to a detailed summary of retraining programs and costs, and of various allowances for associated room, board, and transportation expenses. The document included a detailed schedule of dates for the changing of the plant's production line, and for the phasing in of workers to new jobs on the line. One indication of the composition of the workforce was a section specifying that it was management's responsibility to gain the consent of the parents of the female workers to shift their daughters to new responsibilities. The

---

45. Zensen, "Rationalization Policies," in *Chūyō Iinkai Hōkokusho* (Report of the Central Committee) (Tokyo: Zensen, 1991).

emphasis in the document on a common enterprise under paternalistic management, and on bread and butter issues close to the hearts of the workers suggested a basic focus of labor-management negotiations on firm-level issues of employment and production, rather than on industry-wide issues of industrial redirection and new investment. Labor officials described the year-long consultations as difficult and detailed, with long discussions about shaping common goals, and then multiple revisions to meet labor's priorities. That the negotiations concluded without disruption of the production process pleased management, and Zensen officials appeared satisfied that labor's goals were met without need for further dissent.

At another large mogul spinner, labor and management opened discussions on the projected closing of a cotton spinning mill in Nagano in the fall of 1991. The local branch of the firm's union represented labor, as the Industrial Policy Bureau of the Zensen and the Cotton Textile Department monitored the negotiations. Management representatives from the plant and Osaka headquarters had to juggle a number of priorities, for the industry has long faced a shortage of both entry-level and trained workers. Since larger textile firms have been reluctant to hire temporary workers to operate the increasingly sophisticated spinning machinery, management must retain the better workers and terminate the less efficient as far as possible, but at the same time must carefully maintain their image of benevolence despite the necessity of breaking the precedent of permanent employment. The mogul firm would not likely risk eroding the harmony of labor relations with some 8,000 workers in the same union in efforts to gain advantages in terminations at any one mill. Generous early retirement benefits and large termination bonuses could be offered to achieve reductions and appease the union, but I was told that such programs at other firms in the past drew off the best workers, namely, those with more experience and more initiative. Management thus looked to retraining and reassignment as a positive alternative, insofar as opportunities were available at other mills or in new product lines. The shrinking national pool of Japanese workers, the priority of retaining the better workers, hopes of maintaining a strong company identification, and the diversification of the firm into new production lines all contributed to the priority on retraining programs and transfers.

Six months of intense negotiations concluded in the spring of 1992 with an agreement on transfers and terminations. Union officials felt they had made the best of a bad situation in negotiating the mill closing, for the agreement stipulated that all workers who wished to be retained in other positions within the firm were promised opportunities. Reassignment would include a period of two months or more for retraining, and also relocation to another plant or mill. Workers would shift their occupational specialty and their residence, but were assured of maintaining their previous level of wages, seniority, and benefits. The mogul spinner also offered generous settlements for those who would depart the firm, and those workers would probably find other work within the area through state-funded retraining or reemployment programs.

I had the opportunity to review the text of the tentative agreement and was surprised to find a two page summary on the industry's future appended to the document. The local union with obvious input from the Zensen offered directions for future investment to maintain competitiveness in textile production, urging investment in new technologies and new products. Management responded with a well-considered reply showing interest and a spirit of cooperation, but had no legal obligation to pay any attention whatsoever to labor's ideas on investment. Labor is neither represented on the board of the firm, nor is there any formal consultation with labor on investment policy, but management does need the cooperation of labor for further adjustments of production and plants in the future. Labor gains recognition rather with reactive voice, for if the workers oppose company policy, they gain the attention of executives interested in labor harmony, particularly in sensitive periods of adjustment.

Patterns of dissent yet cooperation among capital and labor continue to the very closing of a mill. Capital and labor have a common interest in the survival and growth of the firm. Worker identification with the firm rather than with a trade, reinforced by the precedent of permanent employment, contributes to labor's interest in the survival of the firm. Retooling a mill for another product line permits the firm to keep its promise of permanent employment and permits the workers to maintain their employment and benefits, although in another specialty. The closing of a mill presents more of a challenge. Firms in declining industries often cannot afford to retool, particularly if they are diversifying into nontextile areas. Retraining and relocation of textile workers for work in nontextile production lines fosters a strong corporate identity among labor, but demands a great deal of the workers, and also strains the resources of firms that might otherwise reduce costs by bringing in new employees already trained in the area. Thus we find most firms reducing labor by terminations in the process of closing mills. Capital even in this process must retain the good will of labor by providing generous termination allowances; labor, of course, has pressed for the best termination benefits possible. Unions must finally legitimize the process of undoing the promise of permanent employment and recognize management's good faith by accepting the negotiated package of termination benefits and closing the mill peacefully.

The profile of labor relations is the final piece in the puzzle of tripartite interest competition in the textile transition, completing our study of the maverick dynamic. The study of labor's role in adjustment clarifies the place of workers in the tripartite negotiation of change. Labor relations in the textile industry in Japan cannot easily be contrasted or compared with labor relations in the West. Labor did not collude with capital, but neither did it enjoy the autonomy characteristic of labor in the advanced economies of the West.

Studies of labor relations in declining industries offer some parallels with textiles. Nitta Michio's recent study of adjustment in the synthetic-fiber industry provides clear evidence of cooperation and reactive voice, and quite apart from textiles, Nitta's earlier study of the steel industry offers strong evidence of cooperation.[46] Mochizuki presents a picture of what I would term "delayed cooperation" from labor in the process of negotiating the privatization of Nippon Telephone and Telegraph. Mike Mochizuki rightly emphasizes the weakness of neocorporatist linkages between the levels of firm, federation, and national center.[47] I also find evidence for cooperation in Kume Ikuo's brief study of restructuring in the coal industry, although he would argue that labor enjoyed full corporatist partnership with capital and the state.[48] In addition to sectoral case studies, the coopera-

46. "Business Diversification and Human Resource Management Strategy in the Japanese Chemical Textile Industry," *Occasional Papers in Labor Problems and Social Policy* (University of Tokyo, Institute of Social Science), no. 10 (March 1991), and "Conflict Resolution in the Steel Industry," 233–47.

47. Mike Mochizuki, "Public Sector Labor and the Privatization Challenge: The Railway and Telecommunications Unions," in *Political Dynamics in Contemporary Japan*, ed. Gary D. Allinson and Sone Yasunori (Ithaca: Cornell University Press, 1993), 181–99.

48. Kume, "Tale of Twin Industries," 158–80.

tion thesis appears to explain how Japan enjoys relatively peaceful industrial relations although Japanese workers enjoy wages and benefits comparable to those found in other advanced industrial nations.[49] The cooperation hypothesis also draws us beyond the anomaly of "enterprise unionism" with its apparent contradiction between firm loyalty and union interests and offers a way past the antinomies of "corporatism without labor" and full corporatist partnership.

A cooperation model also helps explain labor's citizenship in Japan's changing political economy. Permanent employment and intense company loyalty constrain Hirschman's alternatives of voice and exit as means of addressing discontent within firms.[50] Workers usually do not express their dissent by leaving the mill. The absence of the exit alternative gives greater importance to voice, but also limits voice with considerations such as the survival and growth of the firm. Industrial citizenship for the worker at the firm parallels the citizenship of capital within industry associations. One might draw a parallel also between citizenship in the enterprise union and in the Zensen, where again, the textile workers have no viable alternative to participation in the Zensen. The law does not mandate participation in either the enterprise union or in the federation, yet both appear mandatory in practice.

Aside from labor's role, the study also extends our understanding of "encompassing" organizations in Japan. Labor's emphasis on a group interest, whether the enterprise itself or the industry, has been promoted since the oil shock of 1971 by a wider shift in labor in Japan from the politics of confrontation to the politics of consensus.[51] Labor now cooperates with capital and the state to maintain real wages and hold down inflation, rather than simply demand wage increases. Labor joins with capital and the state to ensure job security and job

49. Taira and Levine, "Japan's Industrial Relations"; Japan Institute of Labour, *Japanese Working Life Profile*. Wages in the textile industry are generally lower than in other manufacturing industries. Textile wages in Japan average close to 70 percent of average wages in manufacturing. In France, Britain, the United States, and other advanced industrialized economies, the textile workers average less than one-half the average wages in manufacturing. ILO, *General Report*, 93.

50. Albert O. Hirschman, *Exit, Voice, and Loyalty: Responses to Decline in Firms, Organizations, and States* (Cambridge: Harvard University Press, 1970).

51. Kume Ikuo, "Changing Relations among the Government, Labor, and Business in Japan after the Oil Crisis," *International Organization* 42 (Autumn 1988): 659–87.

opportunity, rather than press demands on wages that would under-mine the industry or the firm. Macroeconomic priorities have thus supported more encompassing interests among organized labor. The state has responded to a more cooperative attitude among labor with the security net documented above. Thus labor can point to tangible benefits from their shift to consensual politics, but apart from wider international and national economic challenges and state programs, the roots of labor's emphasis on encompassing rather than "special" interests can be traced to the enterprise union itself. Loyalty to the firm pushes labor's perspective beyond a craft or trade to the enter-prise on the one hand, and to survival of the industry on the other.

The enterprise base of organized labor also helps explain Japanese corporatism from the perspective of industrial relations. The enter-prise remains the focus of industrial relations, the center of wage bar-gaining and joint consultation, and the keystone of the structure of labor organization. The link between loyalty to the firm and union interest can be explained by the model of cooperation, a second fea-ture of this corporatist strategy of change. This corporatism is "soci-etal" rather than "state corporatism," but with emphasis on the interaction of state and labor in the development of a tripartite frame of negotiation. Inagami has argued that the consolidation of labor's voice was largely due to labor's own efforts and was thus more decid-edly "societal" than forms of societal or "neo-corporatism" found in other societies.[52] He goes on to suggest a blend of neoliberal trends within neocorporatist patterns of industrial relations. His point deserves attention, but he overlooks how state bureaucracies pro-moted labor federations and how capital drew organized labor into the information and negotiation circles. Consolidation of organization and interest provide evidence of societal corporatism in the inclusion of labor's reactive voice in Japan, though only with the consent of state and capital.

52. Inagami, "On Japanese-Style Neo-Corporatism."

7 The General Trading Companies

Since mutual benefit is the key to shared interests, adjustment policy must bridge private and public interests without eroding the advantage of either. Joint efforts at restructuring must meld the special interests of union, firm, or sector with the broader goals of federation, association, or industry. Merchants have traditionally played a central role in brokering the diverse interests of capitalist society, linking consumers to suppliers in the common enterprise of a sale in which maker, merchant, and consumer can all benefit. Merchants complete our story of the textile transition, for makers need merchants to help moderate decline and maintain hierarchy, and makers need merchants with whom they can find common interests and ensure joint profits.

The General Trading Companies (*sōgō shōsha*, or GTCs) have also played a part in the textile transition; they have sustained continuity, facilitated change, and extended the goals of firms to include interfirm and industry interests. Traders import raw materials for the mills at moderate prices, but they also import the cheaper textiles, which has doomed many of those same mills. Traders create demand for the higher value-added products of the adjusting mills, but also import fashions from Europe and the United States that challenge the local products. The term "trade" usually denotes foreign commerce in exports and imports. The GTCs handle most of Japan's imports and exports and also direct the local commerce in textiles among producers, wholesalers, and retailers. The term "trade" also includes purchases and sales. The GTCs link suppliers to consumers through consignment sales or direct purchase and resale and also finance and coordinate production in the midstream sector of the textile industry. M. Y. Yoshino and Thomas B. Lifson describe GTCs as "coordinators of product systems" and chart the extension of the traders' role into various stages of production, and into multiple functions within the overall enterprise.[1] A recent survey of GTC functions in the textile industry has similarly highlighted their systematic "organizing function," importing raw materials for spinners and synthetic-fiber makers alike, and then purchasing the yarn and fiber of those same customers.[2] They also finance production, oversee transport, and market goods as necessary. All mills depend on the information, marketing expertise, and networks of the traders, and smaller producers and brokers depend also on the houses for financing and basic materials for sewing and weaving. Prominence in both procurement from producers and consignment to merchants gives them extensive leverage for monitoring and managing markets in the industry.

The market function links both organizations and interests. Under capitalism producers and consumers are linked by either the market or a bureaucracy. Traders stand between the market and the bureaucracy and, at least among the major mills, help sustain patterns of dominance in the industry through long-term alliances with the larger producers. Neither the state nor any one firm commands sufficient resources to dominate the process of textile production from fabrication to wholesaling in Japan. Yet neither is the market permitted free play in a declining industry. The trader acts as a buffer for risk among producers confronted with destabilizing market fluctuations in the process of adjustment. For instance, the mills benefit from a reliable supply of raw materials at reasonable prices and large-scale, consistent purchases of their products. Stability is a further benefit for the producers, but producers who enjoy the advantages must also abide by limits on their profits.

Recognition of limits demands discipline, and traders can bring discipline to an unruly market. Mark Fruin cites discipline as a general feature of interfirm relations in Japan, where a breach of "acceptable business practices" often results in ostracism from the network, "a modern form of village ostracism [*mura hachibu*]."[3] Boundaries among Japanese firms may be less distinct than in the West, but there is little ambiguity about the limits of special interests or the necessity

1. M. Y. Yoshino and Thomas B. Lifson, *Japan's Sogo Shosha and the Organization of Trade* (Cambridge: MIT Press, 1986), 43.

2. "General Trading Houses: Borderless Trade Age Is Coming," *JTN*, May 1990, 53.

3. Mark W. Fruin, *The Japanese Enterprise System: Competitive Strategies and Cooperative Structures* (Oxford: Clarendon Press, 1992), 10.

for cooperation in comprehensive or encompassing interests. Cooperation in interfirm efforts demands procedure and conformity to precedent. Yoshino and Lifson link discipline to interests: "For every client there is a constant tension between maximizing its own welfare and maximizing the welfare of the system. The *Sogo Shosha* [GTC] acts as interpreter of system welfare and also attempts to act as enforcer of system discipline."[4] Mavericks and dissenting moguls diverge from the compass of adjustment, but not from the precedents of bounded competition with the trading houses. Whether in procurement of raw materials or in joint-ventures abroad, the mills worked hand in hand with the trading houses.

Interest negotiation in the adjustment process has taken various forms: the state coordinates; the member firms of an association compromise; labor cooperates. The mediation of markets by the trading houses for the textile makers is a further example of coordination, and one that helps unravel those deeply rooted patterns of interfirm ties in the industry which sustain hierarchy and moderate decline across the decades of adjustment. But the traders tell us of more than the direction of adjustment, for we also learn how these uniquely Japanese establishments help stretch the limited interests of firms into the wider goals of joint ventures. Trading houses establish precedents and networks of commerce and production that encourage large-scale enterprise, and establish long-term relations that help sustain cohesion and identity among the makers.

Interests

The moguls and the mavericks are the upstream, concentrated sector of Japan's textile industry, but concentration remains atypical in an industry characterized by both specialization and small-scale production at multiple independent enterprises. The bewildering diversity of organizations and interests in Japan's textile industry has caught the attention of foreign observers: "The structure of the Japanese industry is one of extreme deverticalization, with a few very large fiber-makers and spinners at one end, and a multitude of smaller and medium-scale enterprises specialized in weaving (under subcontract

4. Yoshino and Lifson, *Japan's Sogo Shosha*, 47.

with big spinners of General Trading companies) on the other. The latter are specialized both in different processes of weaving and dyeing,
and in different varieties of textiles."[5] Dispersion of expertise and
machinery to multiple smaller and midsize firms in midstream and
downstream areas of textiles and clothing such as weaving, dyeing,
finishing, and garment making leaves the traders with the task of
coordinating the industry by brokering interests across a wide spectrum of firms. Unlike producers in Brazil, Germany, and Italy, Japanese textile firms can call on the skills and resources of not only
smaller brokers, but also the huge trading houses of the earlier
zaibatsu, now linked with the *keiretsu* groupings and closely allied to
the group's bank. Even small, specialized producers need financing,
technology, information on product demand, and access to marketing
networks. The trading houses can fill all these needs and make a profit
on the transaction.

Trading houses play a major role among the more integrated and
concentrated upstream sectors of spinning and synthetic-fiber production, where there has been but little success in efforts to integrate
spinning, weaving, and finishing within a single firm. Nor do mills
direct their output to an intrafirm network of wholesale and retail
outlets. The extensive role of the major mills in domestic and international textile markets might well demand a huge investment of human
and capital resources in marketing and in the management of procurement and sales; however, the mills make a relatively modest
investment in these areas as they rely rather on the information, expertise, and supply and marketing networks of the traders and concentrate their resources instead on controlling production costs,
upgrading production techniques, and designing products for new
competitive niches. The mills focus on manufacture and the merchants
on markets; this supports the shared interests of maker and trader
without eroding the advantage or diluting the expertise of either. The
specialized structure of producer firms and dispersion of resources

5. OECD, *Structural Problems and Policies in OECD Countries: Textile and
Clothing Industries* (Paris: OECD, 1983), 142. An ILO report a decade later repeats
the same theme. "Deverticalisation is a feature of the textile industry in some industri-
alised and high-income developing countries such as Brazil, Germany, Italy, and Japan,
where thousands of small specialised firms maintain close relations with each other and
with trading companies." ILO, "Working Conditions in the Textiles Industry in the
Light of Technological Changes," in *ILO Sectoral Activities Programme, Textiles
Committee, Twelfth Session, Report II* (Geneva: ILO, 1991), 11.

across the wider industry invites the kind of coordinating role that only the large-scale trading houses can play.

Major mills and trading houses grew up together across a century of modern textile manufacture in Japan. Trading houses have been procuring raw cotton and machinery and exporting yarn and fabric for the spinners from the outset. The traders quickly diversified into multiple areas of commerce and manufacturing and have grown far larger than the mills in the past few decades. Among trading houses with close links to the moguls and mavericks, I have selected four with strong interests in textiles. Leaders among the trading houses in sales, Itohchu reported total sales of $146 billion and Marubeni, $134 billion, in 1990. Tomen and Nichimen ranked seventh and eighth among the GTCs, with total sales at Tomen in 1990 of $48 billion and Nichimen of $43 billion. The scale of sales in textiles alone at these giants puts them among the leading firms in the textile industry, but textiles have accounted recently for only about 10 percent of their yearly sales. Among our selected producers, Toray with sales in textiles amounting to about $3.25 billion in the same period is the closest competitor in scale of sales to the smaller of the major traders. Toyobo, Nisshinbo, and Tsuzuki Spinning reported significantly smaller sales than the GTCs.

The trading houses are active in a variety of roles in the upstream sectors of spinning and synthetics fibers.[6] Traders supply the raw materials, supply weavers and fabric makers with yarn or filament from the major mills, and sell mill products at home and abroad. Itohchu, Marubeni, and the others join the mills in developing production areas at home and abroad and jointly promote enterprises in both areas. The trading houses also help the mills form marketing strategies by sharing information on production, supply, and sales at home and abroad. Indeed, only large-scale suppliers can serve the huge demand at the major mills, which are reluctant to warehouse more than two months' supply of raw cotton. The volume and vari-

6. Tanaka Susumu, "Shuyaku ga kōtaishi shijō ga kaette mo mirai wa akarui—sōgō shōsha kara mita sen'i bisyon" (Though the functions shift and the market changes, the future is bright—a textile vision from the perspective of the general trading company), *Kassen Geppō*, April 1989, 25–33. The Japanese term for "major cotton spinner" is *ōte bōseki*. Much of this information was provided in an interview with officials at Itohchu, Marubeni, Nichimen, and Tomen in Osaka and Jakarta.

ety of products from the major mills can only be marketed by large-scale brokers. Priorities of scale or volume, reliability of supply and demand, and specialized expertise link the major mills with the leading trading houses. Shared priorities support the GTC role of coordination, which in turn fosters continuity in ties between the major mills and these huge merchants.

A trader will, for instance, purchase yarn from the spinners and contract it out for dyeing.[7] The spinner is paid in full by the trading company, and the trader arranges for the yarn to be transported directly from spinning plant to dyeing plant. Upon completion of the dyeing process, the trader conveys the yarn to a first-level wholesaler to market the dyed yarn. Traders will often serve as go-betweens with the raw-yarn makers and the weavers or knitters, even providing credit for the latter to purchase equipment, and may then broker the subcontracting of weaving or knitting for the spinner. Once the spinner receives the woven or knitted fabrics, he may then call on the trader to market the goods. But the traders can play a more independent role in production by simply purchasing the raw yarn from the producers, and then contracting out the weaving and knitting process, and later the dyeing process, before passing the goods on to a second-level wholesaler or even on to garment makers. One finds a remarkable combination of financial leverage and marketing expertise at the trading houses sustaining efforts at market coordination. In the absence of integrated producers capable of in-house coordination of the entire line of manufacture, processing, and marketing, the trading companies fill the vacuum by moving goods from mill to market.

Differences in structure and performance between mill and trading house draw attention not only to the need for coordination, but also to the necessary resources. A GTC can marshal a unique range of resources to coordinate supply and demand. Close ties within distinct groups of companies is one key to the effectiveness of coordination among the trading houses. Although most of the major mills remain relatively independent of *keiretsu* ties, traders stand together with banks at the very center of these huge commercial, industrial, and

7. Business Intercommunications, *Distribution Systems in Japan* (Tokyo: Business Intercommunications, 1985), 249, 257.

Table 6. Selected general trading companies: Structure, 1990
(assets expressed in billions of U.S. dollars)

Firm	Employees	Assets	Debt/Total Assets
Itohchu	7,262 (−.09)[a]	58 (+2.0)	.59
Marubeni	7,364 (−.08)	66 (+2.5)	.61
Tomen	3,564 (+.09)	18 (+1.4)	.58
Nichimen	2,864 (−.14)	16 (+1.8)	.60

Sources: Please see the source note to table 5. The ratios of debt to assets are explained in the source note to table 4.
[a]Ratios in parenthesis represent comparison with a base of fiscal year 1980. Employment at Itohchu, for instance, declined 9 percent from the base year total of 8,036 workers in 1980, to 7,262 workers in 1990. Assets increased 200 percent from a level of about $19 billion in 1980 to $58 billion in 1990.

financial alliances.[8] Financing is critical for producers and traders alike. The trading houses are generally more highly leveraged than the mills, reporting debt to total asset ratios in March of 1988 at about 60 percent (see Table 6).[9] Insurance companies and banks retain the major shares in both mills and the trading houses, but the banks play a much larger role in the business of the trading houses. Commercial banks provide the trading houses with the credit necessary to finance coordination of production and marketing among the dispersed textile industry, for the banks have neither the information, market expertise, nor the experience in the textile industry to monitor and profitably invest in coordinating the production process. The traders in turn serve as "banks" on a smaller scale, providing financing for producers and retailers alike, enabling the banks to profit with the growing scale of GTC business, including daily transactions in letters of credit and for-

8. The term *keiretsu* refers to groupings of larger firms in Japan distinguished by extensive intercorporate transactions, and often consultations on intercorporate commercial and industrial policies. Three features of the keiretsu suggest organization and procedures quite different from the centralized conglomerates in the United States: (1) a sense of community and mutual shareholding; (2) specialization of the firms in specific industrial and commercial sectors, and hierarchy among them according to size; and (3) flexibility of relations among member firms. See Rodney Clark, *The Japanese Company* (New Haven: Yale University Press, 1979). Itohchu belongs to the Dai-ichi Kangyo Bank Group, Marubeni to the Fuyo Group. Tomen has ties with the Tokai Bank Group, and Nichimen with the Sanwa Bank Group. See Toyo Keizai, *'92 Kigyō keiretsu sōran* (An overview of keiretsu enterprise, 1992) (Tokyo: Toyo Keizai, 1992), 447–48.
9. See reports on Itohchu, Marubeni, Tomen, and Nichimen in Diamond Publishing, *Diamond's Japan Business Director 1989* (Tokyo: Diamond Publishing, 1989).

Table 7. Selected general trading companies: Performance, 1990
(in billions of U.S. dollars)

Firm	Sales	Profits
Itohchu	146 (+.92)[a]	0.135 (+3.2)
Marubeni	134 (+.86)	0.121 (+.73)
Tomen	48 (+1.0)	0.042 (+2.4)
Nichimen	43 (+1.3)	0.036 (+2.1)

Sources: Please see the source note to table 5.

[a]Ratios in parenthesis represent comparison with a base of fiscal year 1980. For instance, sales at Itohchu increased by 92 percent over the base year total of $75.8 billion to $146 billion.

eign exchange, but also with the growing dividends from profitable GTC investments in smaller firms. Itohchu, Marubeni, and others thus can finance coordination efforts with resources unavailable to mills with ambitions of marketing their own products.

Discrepancies in structure and function evident between producers and traders have contributed to differences in performance over the past decade. Interfirm interests have hindered neither growth nor adjustment. The trading houses not only sell more textiles than the producers, they also report higher growth in sales and profits across the past decade, as evident in table 7. Apart from Tsuzuki Spinning, which has aggressively expanded capacity across the decade, the selected textile firms reported a growth rate in sales of between 10 and 46 percent. The GTCs reported growth in sales of between 86 and 130 percent. Traders also report generally higher growth in net profits. Neither traders nor mills have found strong growth in sales and profits in the textile industry. Annual sales of all products have better than tripled at the four trading houses since the late 1970s, but not in sales of textiles. In fact, the share of textiles and fibers in the annual net sales of the GTCs has slipped from about 20 percent in 1978 to 10 percent in 1990.[10] Although textiles have not kept pace with growth in chemicals, construction, and foodstuffs, it is clear that textiles remain an important area of GTC investment and activity.

10. Dodwell Marketing Consultants, *Industrial Groupings in Japan, 1978* (Tokyo: Dodwell Marketing Consultants, 1978), 363.

Whatever the differences between mill and merchant, they find common ground for joint efforts. Manufacturers retain one trading house as their major supplier and purchaser, and two or more other houses with smaller contracts to lessen their dependence on their main partner. In interviews at the mills, I found spinners resigned to the close relationship with the traders, but anxious about links between the same houses and their competitors among the spinners. Toray cited Itohchu, Mitsui, and others as its main suppliers and purchasers. Toyobo and Nisshinbo listed Itohchu, Marubeni, and others. Tsuzuki Spinning, however, listed Itohchu and a local Nagoya trader, Toyoshima, as its main suppliers and purchasers.[11] Apparently, the local traders in Nagoya were more flexible and responsive to Tsuzuki's needs, but only the leading Osaka-based GTCs could command the necessary capacity to supply large lots of raw cotton to or purchase large lots of yarn from Tsuzuki. That Tsuzuki directs some of its business to the Osaka traders despite the proximity of the smaller Toyoshima trading firm suggests a linkage of scale between the larger mills and the major trading houses. It also suggests the discipline that GTCs bring to the volatile market of raw cotton. Itohchu, Marubeni, Tomen, Nichimen, and the other major houses can better respond to the large-scale supply and marketing needs of the big mills than to those of the smaller, specialized trading firms. At the same time, the major trading houses enjoy economies of scale in the larger transactions with the moguls and mavericks.

Coordination at the trading houses supports continuity of supply and demand for the mills, which in turn reinforces the hierarchy of the moguls and their survival. Structural demands for scale, reliability, information, and expertise bring mills and merchants together in a common enterprise. The community of interests among mill and merchant is supported by their distinctive paths of development as manufacturers and marketers. Moderation of market fluctuations provides the cement for their cooperation. Traders and producers divide the market risk in procurement, production, and sales. Just as commercial banks can reduce the risks of exposure by lending capital through the trading houses and their textile divisions, which are specialized and well-informed about credit-worthiness of smaller producers and mar-

11. Nihon Sen'i Kyōkai, ed., *Sen'i nenkan 1990* (Textile yearbook, 1990) (Tokyo: Nihon Sen'i Shimbunsha, 1990), 354, 383, 385, 333.

keters, so also can the trading houses and producers together spread the risk of market fluctuations by collective action in procurement of raw materials and sales. I would point to a comparable benefit of sharing expertise. Collective action brings to the market the specialized talents and information of both producer, and financier/marketer. Structures and interests here support cooperation in the GTC role of coordinating markets. Coordination in turn reinforces the differentiated but complementary strengths of mill and merchant and fosters extension of their firm-specific interests to an interfirm interest.

Anomalies

In textile production a pattern of dispersion or disintegration downstream and concentration upstream can be found in many countries, but still other anomalies appear distinctive to Japan. For instance, Japan remains one of the leading cotton spinners in the world, without any local source of raw cotton. The expensive, automated technology found in Japanese mills stays in production around the clock. Japanese spinners procure supplies of top quality raw cotton at the most competitive world prices, despite purchasing almost entirely through Japanese trading companies. Spinners can rely on a reliable and regular supply of high-quality cotton. They compound the anomaly of cotton manufacture without local cotton by purchasing exclusively from Japanese GTCs. Long-term relations with Itohchu or Marubeni or others leave the spinner tied to one or two trading houses, but assured of reliable supplies at a moderate price. Again, coordination leads to continuity, based on a common interest in reliable supplies, orderly markets, and wide access to the best cotton varieties available.

A cooperative umbrella of purchasing procedures at fair prices with reliable supplies is particularly important for an industry totally dependent on imports of raw cotton. The advantages of purchase through the GTCs also include a broad range of cotton varieties and qualities for selection by the individual spinner. Mills and traders share the risk in the purchase of raw cotton. The trader buys in advance at relatively lower prices on a soft market on the supposition that he can recoup a profit by delivering the cotton to a mill later when prices have gone up. Profitable advance purchases demand capital and intimate knowledge of cotton suppliers, pricing, and the projected needs of consumers. Seki Keizo cited three factors determining the basis for the actual delivery price to the Japanese mill of "futures" purchased in advance on the New York market:[12]

1. The New York futures quotation for a specified month of delivery
2. The difference in price between the futures quotation and the spot price for a specified grade and staple of cotton.
3. The freight charges, insurance fees, interest payment and other incidental costs incurred in getting the cotton to Japan.

Long-term relations between larger mills and major trading houses further complicate purchase and sale of raw cotton. Only the major trading houses can provide the volume and variety of raw cotton necessary for both mass and specialized production. The big mills, however, represent the most profitable and consistent customers for the trading houses. Good relationships take precedence over windfall profits, which ensures that neither party will take undue advantage of the other, despite multiple opportunities in the volatile cotton futures market.

Traders buy and sell cotton on the basis of common rules approved by the Liverpool Cotton Association, which establishes a framework of legal accountability between traders and mills. David Morris argues that the advantage of accountability persuades most mills to purchase through traders rather than directly from the growers and adds that fifteen major trading companies control nearly 90 percent of the cotton trade across the world.[13] But few major mills in other advanced, indus-

12. *A Study of the Japanese Cotton Industry: Its Past, Present, and Future* (Tokyo: Tokyo University Press, 1954), 118–19. I heard similar explanations in interviews at the major trading houses in Osaka.

13. *Cotton to 1993, Fighting for the Fibre Market*, Special Report #1151 (London: Economist Intelligence Unit, 1988), 40. An ILO report offered the following assessment of the trade in cotton. "International marketing is, however, dominated by 15 cotton traders, comprising three distinct groups: two European corporations (Ralli Brothers, United Kingdom; and Volkart Brothers, Switzerland), eight U.S. companies (including Cargill, and Bunge and Born), and five Japanese trading houses (Tomen, Sumitomo, Itohchu, Marubeni, and Nichimen). In cotton, as in other agricultural commodities and metals, the traders dominate world marketing, and their influence extends through their formal and informal links with the transnational banks and in some cases through backward (plantations and ginning) and forward (spinning and weaving) integration. Some traders, most notably the Japanese trading houses, have become active in the international buying of clothing as well." United Nations Centre on Transnational Corporations (UNCTC), *Transnational Corporations in the Manmade Fibre, Textile and Clothing Industries* (New York: United Nations, 1987), 15.

trialized nations remain so closely tied to the trading houses of their own nation. Itohchu and Tomen stand among the larger cotton traders, with Marubeni and Nichimen not far behind.[14] Japanese spinners continue to purchase cotton almost exclusively from the Japanese trading houses. Seki argues that "the existence of powerful trading companies with a world-wide trade network, in combination with the business ability of the spinning companies supported by large operating funds, enabled the industry to import its raw cotton cheaply."[15] Trading companies certainly do extend their purchases far and wide across the globe in search of better prices and better qualities of cotton.

How can we account for the persistence of such close interfirm cooperation? Major cotton suppliers from across the world now have offices close to the headquarters of Toyobo, Kanebo, and Unitika in Osaka. We find agents for U.S. cotton shippers in Osaka processing orders from the trading firms rather than directly from the spinners themselves, yet contacting mills directly to advertise types and volumes of cotton available. The anomalies continue. Mogul and maverick firms now have their own offices in the cotton growing areas of the United States, with agents dispatched directly to the cotton fields to determine quality of the crops and available volumes. Why don't these offices arrange purchases directly with U.S. shippers? People in the industry cited three reasons for the persistence of purchases through the general trading companies of Japan: access, cost, and reliability. Common interests are built on shared advantage, and the advantages lead millers to maintain exclusive purchases through the GTCs.

Certainly there are advantages, for the very scope of the GTC purchasing network would catch the attention of any customer. Trading companies can penetrate distant markets and remote hinterlands in search of better crops and prices, whereas the spinners can only operate in a few major and relatively well-organized cotton markets such as that of the United States. Indeed, the Japanese spinners have yet to extend their offices to the cotton-producing areas of one neighboring nation with a huge cotton crop: the People's Republic of China. An example best explains the cause and consequence of this anomaly.

14. Tomen has recently spun off its raw-cotton division as a wholly owned subsidiary called Toyo Menka. I will refer to this subsidiary by the title of the parent firm.
15. *Study of the Japanese Cotton Industry*, 119.

Both spinners and traders complained about problems of ascertaining prices and availability of the prized Xiangjiang cotton in western China, particularly during the flood season. A trading-house official in Osaka explained that difficulties, dangers, and hardships of travel in the Chinese hinterlands discouraged even GTC agents. At the same time the lack of reliable government statistics or other information on the cotton crop in China leaves traders dependent on their own agents, who eventually do get to the sites and provide information to the home office. But if the traders must struggle to obtain reliable information on the Chinese market, the spinners alone have no access whatsoever. Without agents residing in China, the spinners must depend on the trading houses to obtain information and prices of the Chinese cotton.

Cost is a second and very compelling reason to maintain the long-established precedent of procurement through the trading houses. Spinners are generally quite satisfied with the cost of the cotton they obtain, in comparison with the prices they would pay U.S. shippers directly. And beyond the relatively accessible U.S. market, the spinners become more dependent on links with the traders in cotton-growing countries providing less information on supplies. The GTCs can buy on markets across the world and bring to the market experience and expertise in pricing and quality of cotton. Mills train a few cotton classers among their staff, but would need to devote considerable resources to develop comparable expertise in cotton purchasing.

Reliability is a third factor supporting exclusive purchases through trading houses, because in part of the vulnerability of cotton plants to climatic changes, contamination, and damage in transport. Problems of nurture and harvest of the delicate crop can easily disrupt the supply line, and leave shippers exposed to wild fluctuations in the price of cotton. If the price of cotton goes up between contract signing and actual delivery, the shipper may not be able to cover the margin nor fulfill the contract, forcing mills without cotton to shut down. Executives at major trading houses explained the role of the GTC in insuring reliability. GTCs provide a principal to principal forward shipment business, with six months' forward delivery. If the shippers do not fulfill the contract, on the basis of which the GTC has already sold the cotton the Japanese mill, then the GTC takes the loss, not the Japanese mill. The GTC will also take the loss if the Japanese mill will not accept the cotton delivered, due to water or insect damage.[16] As the Japanese spinners upgrade their machinery and spin higher counts of cotton, they demand finer varieties of cotton and consistent quality of supply. Long-term relationships between moguls and trading houses ensure both. A history of ties between individuals in the raw-cotton sections of both spinner and trader provides a foundation of personal relations to support the written contract. Since spinners focus more on quality and specialized production, they will not likely be willing to sacrifice this reliability on behalf of marginally cheaper cotton obtained directly from the shippers.

Cost, accessibility to world markets, and reliability of shipments help account for the exclusive purchase of raw cotton through the trading houses. The mills may supplement supplies or purchase certain special cotton varieties through the medium-size or smaller traders, but still purchase the bulk of their cotton through the major GTCs. Factors of cost, access, reliability, and scale ensure the relational ties beyond contract that bind moguls and mavericks with the major GTCs. One could cite also the multiple linkages between producer and trader across various divisions of both firms, including imports, exports, offshore production, and domestic production and commerce, which reinforce long-term relations. Such ties help sustain hierarchy among the producers, but do not appear to discourage the rise of a few mavericks.

Rational motivations of cost, access, and reliability link the GTC role of coordination with the result of continuity among the moguls and mavericks. The explanation helps unravel embedded relations between makers and merchants in the textile transition. We discover not just anomalies such as a reliable supply of raw cotton despite the absence of a local crop, or exclusive purchasing despite an array of possible suppliers. We also find rationality in the anomalies, which explains both their persistence in a market economy and the remarkable growth and survival of the spinning industry. The results suggest two conclusions. First, conditions of supply and demand, now channeled by the makers and merchants in tandem, support extension of special interests between maker and merchant and control of divisive

---

16. I was given these explanations at the raw cotton departments of Itohchu, Tomen, Nichimen, and Marubeni and at the Cotton Traders' Association in Osaka, in the fall of 1991.

interests. There is a market rationale for encompassing interests. Second, precedents now long embedded in organizational structures such as the cotton procurement departments of the GTCs, have developed with their own rationality, apparent in exclusive purchasing contracts with major mills. Anomalies with deep historical roots provide a rationale for seemingly irrational practices in procurement of raw cotton. Embedded anomalies serve to moderate a market inefficiency such as manufacturing without a local source for the necessary raw materials.

Adjustment

Mill and merchant join forces to shape a more encompassing, interfirm interest. Extending interests helps resolve market inefficiencies such as large-scale manufacture in the absence of local sources of raw materials. Our emphasis on combining interests highlights the rationality of apparent anomalies, their persistence and role in sustaining the larger mills in the adjustment process. Mills and merchants have found a common interest in nearly exclusive procurement of raw cotton through Japanese GTCs, something of an anomaly among major cotton manufacturers across the world. The deep historical roots of such cooperation and the long-term link between imports and exports through the same GTC might suggest an historical exception with limited relevance to recent changes in the industry. The further anomaly of joint ventures abroad in the adjustment process, however, parallels these earlier efforts of mill and merchant to extend their interests into a common enterprise and suggests more than a historical curiosity. Cooperation in the procurement of raw cotton has deep historical roots; cooperation in offshore production has developed only in the past few decades.

Among the anomalies of international investment in textiles, the Japanese mills go abroad in tandem with the large trading houses, as opposed to the independent investment abroad of their European or U.S. competitors. The mutual advantages of such cooperation are as follows: "Trading-house involvement has the advantage for independent Japanese companies of reducing problems in host country relations, of providing market knowledge and distribution channels and, of course, of risk reduction and minimizing capital requirements. . . .

For the trading houses, association with manufacturers can entail profits in providing supplies and raw materials necessary for the investment as well as through their role in international marketing."[17] Traders with long experience of commerce in the target area help with financing, marketing, and imports of raw materials and machinery. Japanese mills can contribute their knowledge and expertise in spinning, but with limited capital resources must find a Japanese partner as well as a local partner to put together the extensive capital necessary for start-up costs. Spinners also seek out traders as investment partners to reduce their exposure in case of local political instability or other problems that might close the offshore plant.

Mill and merchants join forces abroad because of a consensus over goals. One can distinguish three goals apparent in strong Japanese investment in textile production in the developing economies of Latin American and Southeast Asia. First, Japanese firms hope to retain a place in these export markets by producing locally and thus avoiding import barriers. Second, the Japanese firms plan to gain a share in the growing export trade of these nations, including a share in export quotas to U.S. and European markets. Third, the Japanese investors look to profit by transfer of technology, gaining a further return on their investment in technological development.[18] A closer examination of investments in Jakarta, however, reveals the difficulty of maintaining such a consensus on goals.

The joint investments often follow upon some years of exporting ties between spinner and trading house. For instance, Itohchu had been handling Toyobo exports to Indonesia for some time before the two firms agreed to invest in production. They joined to found P. T. Unilon Textile Industries in December of 1970 with only 150 spindles. It took a short three years to put in place 11,600 spindles, 350 looms, and basic dyeing equipment. Continuing investment has brought the number of spindles up to 33,500, together with some 700 looms, further capacity for dyeing, and a workforce of 1200. Itohchu and Toyobo each hold a bit less than a third of the shares, and local partners the final third. Yet the top management of the firm includes ten Japanese employees posted from Itohchu and Toyobo, and only three

17. UNCTC, *Transnational Corporations*, 35.

18. ILO, *General Report of the Twelfth Session of the Textiles Committee, 1991: Report I* (Geneva: ILO, 1991), 69; see also UNCTC, *Transnational Corporations*, X, 33.

locals.[19] The initial investment brought together a Japanese mill and a GTC in a strategy to penetrate the Indonesian market from within. Toyobo offset a decline in exports to Indonesia with profits gained from sales within Indonesia of Unilon products, just as Itohchu added a profitable source of foreign investment to their portfolio and extended their presence in a country rich in oil, timber, and other resources. Spinner and trader cooperated to enhance their competitiveness.

Itohchu and Toyobo established Unilon to take advantage also of programs for promotion of import-substitution funded by the Indonesian government. A weak Indonesian domestic market in the early 1980s, local government encouragement of exports, and the availability of Indonesian quotas to lucrative markets abroad turned their attention to exports. Now both Toyobo and Itohchu draw their benefit from Unilon in products rather than directly from dividends. Toyobo imports its share of the production for resale through a Toyobo marketing network, and Itohchu sends its Indonesian textiles to Milan, London, and elsewhere. But production for export markets in Japan and Europe demand more sophisticated technology and more reliable quality control than production for the local Indonesian market. Unilon has aspirations and some experience in integrated production from spinning, through weaving and knitting, and even to the dyeing process. Itohchu, Toyobo, and the local partner have raised the capital to obtain more advanced technology such as air-jet looms from Toyoda, and even a water purification plant to ensure clean water for the dyeing process. Here Toyobo with its long experience in production and expertise in the newest technologies plays a critical role, as Toyobo specialists not only order the machinery and design the production line, but also oversee the manufacturing and initial finishing processes. Unilon and other firms in Indonesia face problems with quality control in areas demanding advanced skills such as weaving, knitting, and especially dyeing. The pattern of cooperation between trader and spinner has taken on a new dimension with efforts to upgrade production for export markets. Toyobo no longer simply runs spinning machines in a new setting, but now must find ways to

19. Toyobo has posted eight employees to the firm, and Itohchu two. Toyobo has responsibility for the presidency, factory management, equipment, spinning, weaving, and processing. An employee from Itohchu headquarters serves as vice-president with responsibility for operations, finance, business, and labor.

maintain quality in a much more diversified production process of goods for Japanese and European consumers. Can the mill and merchants maintain consensus on goals in this transition?

Disparity of interests and resources between mill and merchants have led to new forms of foreign investment. The GTCs, with their greater resources, access to international textile markets, and experience in offshore investment, are active in Indonesia and elsewhere in textile production without the mills. We also find the GTCs investing in Indonesian products but not in production. Certainly production demands resources that procurement does not. For example, Itohchu officials in Jakarta emphasized the difficulty of establishing, maintaining, and upgrading mill capacity. As traders rather than producers, their interest does not lie in development of productive capacity, but rather in integrating a production line among firms and specialties within or beyond national borders. Certainly Itohchu profits from its Indonesian production commitments in joint investments with Toyobo, in P. T. Indonesia Asahi Chemical, and with a Japanese garment maker. Itohchu also puts out short-term contracts to garment makers in Indonesia for specified designs and a certain number of units. Yet it seems logical to assume that if Unilon faces problems with quality control at their own plant, garment makers with materials from other Indonesian plants would face similar and perhaps more severe problems. Whether or not Itohchu succeeds in efforts with local garment makers, the diversity of the trader's efforts contrast sharply with the dedicated focus of Toyobo on production at a single joint venture. Cooperation at P. T. Unilon does not bring together equal partners. Toyobo and Itohchu may jointly contribute expertise and capital to Unilon, but Itohchu has the resources and expertise to exploit multiple opportunities in the Indonesian textile industry. If the Unilon operation falters, Toyobo loses its foothold in the growing textile industry of Indonesia, but Itohchu does not.

A review of offshore investment suggests two conclusions. First, coordination is not collusion. The encompassing interests of mill and merchants may take them abroad in a joint venture, but those interests are neither static nor exclusive. The review of two decades of production at P. T. Unilon suggests a dynamic process in which interests are renegotiated, new types of expertise are demanded, and new resources must be shared. The pattern of cooperation takes on a life of its own in a joint venture abroad which must adapt to changes in the local setting, to opportunities in international markets, and to the changing interests of its Japanese partners. The joint venture in Indonesia neither excludes Itohchu from similar ventures with other Japanese mills and other local partners in Thailand or other nations, nor does it tie Toyobo to Itohchu within the Japan's domestic textile industry or in other investments abroad.

Second, the adjustment pattern of offshore investment promoted in the textile vision aligns with traditional patterns of cooperation between mill and merchant. Again, the embedded character of inter-firm ties within Japanese industry strongly influences the pattern of extending interests beyond the advantages of individual firms even across national borders. The structure of production and marketing in Japan's textile industry and the historical precedent of trading ties with individual GTCs play a major role in determining the pattern of cooperation between mill and merchant in offshore investment. Structure and precedent shape not only continuity of interfirm ties, but also change in investment strategy. Opportunities for shared risk and shared profits between the moguls and the major GTCs encourage continuity of ties, just as the availability of such opportunities for mills accustomed to local production but uncertain of production abroad encourage change. The specialized expertise of mill and merchant and the historical precedents of cooperation in offshore markets eases the path for adjustment in the common compass of change. There was rationality in the apparent anomaly of cotton procurement; there is also rationality in the apparent anomaly of offshore investment only in tandem with GTCs.

Major trading houses in Japan moderate market dynamics for the moguls and mavericks; this is evident in how they procure raw materials and in how they enter into joint investments with the mills in production offshore. The role of coordination brings discipline and rationality to a dispersed process of production and to an array of unruly markets. Coordination permits specialization and differentiation of functions at mill and trading house, fostering interdependence in the pursuit of profit and expanded markets. Coordination also contributes to the persistence of hierarchy among moguls and mavericks through long-term linkages of procurement, marketing, and investment. And if coordination promotes continuity, the example of offshore investment suggests it also contributes to change or adjustment, permitting survival and even prosperity among the mills. But coordination at Itohchu and other houses tells us of more than adjustment. It sheds light also on the intricate process of extending firm-specific interests into the broader interest of a sector or industry. Indeed, one can find parallels in how the multiple structures of state, capital, and labor shape encompassing interests out of limited ones. Certainly the authoritative coordination of the state supports the commonweal at the expense of the separate interests of sectors within the industry, but a similar dynamic exists for capital and labor, whether in compromise at industry associations or in cooperation at labor federations.

Coordination among the traders draws attention to apparent irrationalities in the structure of trade, whether in the exclusive purchase of raw materials, or the practice of joint investment among Japanese mill and trader in production abroad. These anomalies are embedded in enduring structures, such as the raw-cotton departments at the trading houses, or the joint-ventures abroad. Still, such structures foster shared interests in moderating the price of raw cotton and reducing risk in investments offshore. Anomalies embedded in structure and interests help resolve market inefficiencies such as the absence of a local supply of raw materials and moderate market complexities, such as those involved in developing production facilities abroad. That apparent anomalies bring rationality to market inefficiencies provides evidence of rationality in the extension of interests. Cooperation in encompassing interests can enhance competitiveness.

Accommodation of firm-specific interests among mill and merchant in textiles sheds light on the process of extending interests. The study of GTCs in the textile industry reveals the incremental process of building common goals across diverse firms with specialized expertise and interests. Trading houses play a special role here, in part because of their place within the *keiretsu* network. Attuned to the interests of industrial, financial, and commercial units within a family of firms, trading houses appear ideally suited to broker interests among firms within industries. Interfirm interests between mill and trading house help establish patterns of cotton procurement or offshore investment within a sector. The GTCs provide leadership and resources in the interfirm teamwork among spinners, weavers, finishers, and even garment makers in managing the long production line necessary in the textile industry. Here intersectoral interests are promoted as an industry interest.

A GTC role in extending interest horizons can be further explained by looking to the interplay of structure and interest. Coordination links organizations and interests. Functionally differentiated firms have developed into a highly specialized structure of production and marketing in the Japanese textile industry. Integration of the process falls to the GTCs, which have developed with the resources and expertise to bring mills to markets. Shared interests in profit and expanded markets join merchant to maker. Structural specialization in production on the one hand and in marketing and finance on the other foster common interests and activity. Yet coordination is not collusion. There is some tension between maker and merchant in offshore production. Negotiation of a common, encompassing interest between maker and merchant is supported by common structures such as a joint venture, but also challenged by the changes in the profile of market opportunity at the venture for each.

The study of the coordinating function among the traders helps further clarify the process of accommodating interests in the textile transition. The GTC role also sheds light on the corporatist negotiation of change in the industry. The enduring structuring of interests among the mills in their industry associations depends on the persistence of hierarchy among the major firms and the survival of the producers in a changing market. The coordinating role of the traders fosters continuity in patterns of supply and demand for the producers, provides the necessary variety of raw materials at competitive prices, and even makes possible investment in offshore production without extensive exposure of capital and technology to the uncertainties of foreign investment. Coordination also promotes changes such as investment in production facilities abroad. Such opportunities give substance to the textile vision, help make adjustment feasible, and further encourage the process of corporatist negotiation of change among state, capital, and labor. Corporatist structuring of interests among the upstream producers promoted a common direction of adjustment without precluding dissent and divergence. GTC ties with moguls and mavericks alike permitted opportunities for dissenters. Some diversification of ties between mills and major trading houses and the GTC practice of serving multiple clients helped maintain interdependence without exclusive ties between any single mill and any single trading house. Cooperation between major mills and major trading houses promoted competition among the limited set of major mills and major houses.

Coordination offers further evidence of the societal character of corporatism in the textile industry. Trading houses mediate markets for the mills. Multiple, congruent mediations within society provide a cohesive base of shared structure and interest supporting the persistent structuring of interests evident among the mills. Yet this coordination is only the initial layer in a skein of interfirm relations that bring mills to markets. The trading houses have alerted us to a very significant, and indeed broad level of mediation in which firm-specific goals are extended into interfirm interests, and eventually to sectoral and industry-wide interests. If labor-management relations suggest microcorporatism at the enterprise, and associations and labor federations indicate a structuring of functional interests within a sector, the trading houses turn attention to intersectoral interests within an industry. Market mediations at each of these levels help define a societal corporatism in Japan far more complex than any structuring of interests based largely on a political party with leverage to shape government bureaucracy with corporatist goals.

8　The Fashion Houses

The merchants of fashion complete our story of markets. Fashion dominates the imagination of mill, merchant, and consumer alike. Textiles may begin with raw materials and machinery, but the final product has long before been conceived, designed, and even advertised. The masters of commodity fashions can be found neither at the mill nor at the trading house, but rather at the major apparel wholesalers. Fashion firms in Japan have prospered with the growth of a wealthier and more fashion-conscious consumer population. Drawing yarn, fiber, and fabric from local and foreign mills, they sell apparel to high-volume retailers. Traders first turned our attention to market and inter-industry ties between mills and merchants, for survival and even prosperity at the mills depends on reliable market access. Both traders and the major fashion houses link mill to market, and mill and fashion merchant in turn help sustain market coordination in textiles by Itohchu and others. Fashion houses tell us not only of interfirm cooperation but also of competition and change in the textile transition, and extend our story from textiles to apparel. Coordination of fashion promotes hierarchy and survival among adapting, reliable mills, and presses the extension of interests from sector to industry and ultimately across borders to an international industry.

Fashion draws us more deeply into the textile transition, but away from mills, fibers and fabrics into the volatile world of fashion and apparel markets. The expertise of the fashion houses, their exposure to markets, and their commitment to the industry provides a foil to the mills and traders. Specialization of roles continues in this highly dispersed and differentiated industry. Unlike mills specializing in production, the fashion magnates must bridge manufacture and marketing. Their expertise includes projections about new styles, design, production, and wholesaling. Unlike trading houses, which coordinate production from textiles to apparel, fashion houses coordinate the production and distribution of an apparel line from design to department store. The transfer of this specialized expertise across firms and borders and the coordination of makers and merchants within and beyond Japan poses new problems of adjustment and extension of interests.

Fashion houses are immediately exposed to consumer markets without the insulation of production cartels or capacity reductions found at the mills. Fashion houses also lack the diversity and financial resources of the trading houses. Deeply committed to rapidly changing markets, these apparel giants can survive only by success in seasonal fashions. Market exposure gives a sense of immediacy to the task of adjustment and extension of interests among these coordinators of fashion. Indeed, the commitment of fashion houses to the apparel industry reinforces the sense of immediacy in market exposure. Trading houses have diversified far afield of textiles, and even the venerable mills with nineteenth century roots in spinning have begun investing in non-textile areas. Itohchu, Marubeni, and others range among various products and markets, with textiles accounting for only 10 percent of their annual sales. Major mills project that textiles will represent at best 50 percent of their total sales by the end of the century. Purveyors of consumer fashions rival mills in total assets, but already register larger annual sales in textiles than many mills or even major trading houses. Given this unique commitment to the industry, effective adjustment appears more urgent, and interfirm teamwork and integration of textile interests more critical among the fashion houses.

The chronicle of industrial change returns now to firms, but to firms without the links apparent among the associated mills. The leading apparel wholesalers or ."fashion houses," such as Onward or Renown or Gunze or Wacoal, have neither a cohesive employers' association nor even a sustained informal coalition of fellow wholesalers finding common ground for adjustment strategies. The forces of communal benefit and industry-wide strategies of cooperation help mobilize common efforts for adjustment among the moguls, but the forces of the market prove far more significant among the fashion houses. There is little evidence for corporatist strategies of change between state and capital among wholesalers hustling to shape and serve consumer tastes. Nonetheless, these firms consulted with the state on industry adjustment policies, gained representation in the advisory council shaping directions of change, and themselves began to adjust in line with an industry vision of product specialization and offshore investment. But there are differences. We find greater market exposure and far less mediation between firm and market among these apparel giants. There is more competition and less cooperation than can be found among the upstream producers.

A review of adjustment at the fashion houses poses two questions for our study of neocorporatist styles of restructuring. What role do the fashion houses play in the societal corporatist strategies of change evident at the mills? I suggest here that long-term, stable ties of production and supply with the fashion giants provided a frame or context of reliable market access for the major mills. Moguls could restructure confident of retaining a stable outlet for their textile production, and mavericks could expand confident of further market opportunities fostered by the growing demand for yarns and fabrics at the fashion houses. Links with the fashion houses indirectly supported the blend of direction and dissent at the mills by providing market access for both expanding and contracting spinners. A second question draws attention to a glaring discrepancy between corporatism upstream and more pluralist strategies of change downstream. The fashion houses provided a congenial frame for corporatist strategies of change among producers, but the wide diversity of functions and interests among the fashion houses represented something of a foil to the more cohesive, interfirm corporatist pattern of interest and association among the mills. Why do we not find societal corporatist strategies of adjustment at the marketing end of the industry? The functions and interests of the fashion houses help explain the discrepancy.

Recognition of common interests among firms in an industry give shape and continuity to common policy positions on adjustment and collective action in reduction programs and production cartels. Common interests provide a basis for corporatist styles of cooperation between state and capital. For instance, cooperation in the employers' associations among the mills helped moderate decline during the decades of adjustment, sustain the hierarchy of the mogul firms, yet did not prevent the emergence of a few expanding mavericks. Effective adjustment among the fashion houses allows continuity of supply contracts from the larger mills, in turn supporting survival and persistence of hierarchy. A more immediate question is adjustment among the fashion houses themselves, where the problem of shaping common interests draws us within firms to manufacturing and marketing divisions, and beyond individual firms to intersectoral, and even international integration of an apparel line.

Interests

The argument for fashion houses as both frame and foil for the mills begins with contrasting expertise. Fashion houses develop expertise in shaping and serving consumer demand among retailers, whereas mills find their focus in upstream production for a market mediated by multiple levels of wholesalers. Fashion houses provide market access for major mills, which is critical both for moguls maintaining or curtailing existing production schedules and for mavericks expanding into larger production lots. Fashion houses provide stable market outlets for large-scale producers that can provide yarn, fibers, and fabrics for both large lots of commodity items and smaller lots of specialty items within tight production schedules. Links of scale, quality, and schedule constitute a frame or context of market access for the mills.

Fashion firms do more than provide a market context for corporatist strategies of adjustment, they also provide a foil of noncorporatist adjustment. The diversity of their functions, recent shifts in their production and marketing commitments, and their relatively short history set fashion firms apart from the major mills. There is no parallel among the leading apparel makers of interfirm ties based on similarity of function prominent among the mills. Mills produce yarn and fibers, but fashion houses undertake a variety of tasks from manufacturing to merchandising. A diversity of functions at the downstream fashion houses leads to a diversity of interests among these firms. Conflicting interests in production or marketing, high fashion or commodity fashion, outerwear or inner wear, discourage the formation of cohesive industry associations necessary for corporatist strategies of change. Furthermore, multiple sectoral identifications within the "fashion industry" impede establishment of structured channels of negotiation with other industry associations or with state bureaucracies.

Mills are makers who find common interest with other mills in the JSA, where a community of interest develops around similar functions and shared market priorities. Trading companies coordinate a production process in huge volumes of textiles, moving products to first-level wholesalers. Mills and traders have established a common interest in cotton procurement and offshore investment, where differences are seen as complementary rather than contrasting, mutually supportive rather than divisive. The combined function at fashion houses of apparel maker and merchant complicates the articulation of interests within and beyond the firm. The largest wholesalers of men's suits and women's dresses, Onward and Renown, are listed with the commerce companies on the Tokyo Stock Exchange. The largest manufacturers of men and women's underwear, Gunze and Wacoal, are listed with the textile firms. Yet Gunze and Wacoal market products manufactured outside their firms, and Renown runs various garment-making plants that produce some of the apparel it sells.

Apart from differences, we can also cite continuities of scale and market among the four firms as leaders in production, sales, and market share. These four firms also share a dedication to bridging manufacture and marketing, for fashion leaders must blend the production and wholesaling tasks with networks for both production and sales. Manufacture includes the design of large lots of clothing to clients' specifications, procurement of the necessary fabric and yarn, and final assembly or sewing. Marketing of goods produced within the firm, subcontracted to other manufacturers, or purchased directly from other manufacturers are more traditional merchandising tasks. Makers sell their products to wholesalers, whereas merchants procure their supplies and establish markets. What distinguishes these "fashion firms" is the combination of manufacturing with extensive wholesaling and even retailing within the same firm.

Onward Kashiyama is the largest manufacturer of men's suits in Japan, and a leading manufacturer of women's clothing. Renown developed as a wholesaler and later a manufacturer in the prewar years, and now sews and sells men's suits through its Durban subsidiary, and women's dresses through Renown Look. Gunze developed as a silk-reeler in the late nineteenth century and now has a major share of the men's knitted underwear and women's hosiery markets. Lacking the prewar roots of Onward or Renown, Wacoal began as a garment maker soon after World War II and now ranks among the industry leaders in manufacture and sales of women's foundation garments and lingerie. Despite its origins in manufacture, Wacoal has distinguished itself in sales with its own network of retail stores. A contrast of the fashion houses with the mills and trading houses highlights both market exposure and commitment to textiles.

A profile of the fashion firms can be found in table 8. They are comparable to spinners such as Toyobo and Nisshinbo, yet far smaller than the huge trading houses. Yet unlike either mill or trader, fashion houses devote their personnel and assets almost exclusively to textiles and apparel. A strong growth rate over the past decade at the fashion houses parallels growth at the mills, although growth at the mills includes both textile and nontextile areas. One obvious difference between mills and fashion houses is the low rate of profits at the latter. As the moguls curtailed production and downsized their workforce, Renown and Wacoal added sales staff. Onward and Gunze curtailed manufacturing employment but not sales, as the fashion houses expanded into the apparel market. Renown doubled its workforce across the decade, and Wacoal increased its by 20 percent, which helps explain the drop in profitability at both firms evident in table 9. Sales at the four apparel giants parallel total sales at Nisshinbo and Tsuzuki, but not at Toyobo or the huge Toray. Textiles represent 50 or

Table 8. Selected wholesalers: Structure, 1990
(assets expressed in billions of U.S. dollars)

Firm	Employees	Assets	Debt/Total Assets
Onward Kashiyama	3,258 (−.18)[a]	1.7 (+1.1)	.11
Renown	9,053 (+1.1)	2.4 (+2.3)	.17
Gunze	3,950 (−.36)	1.5 (+1.4)	.24
Wacoal	5,287 (+.18)	1.3 (+.92)	.09

Sources: Please see the source note to table 1. The ratios of debt to total assets are as reported in *Diamond's Japan Business Directory 1991*, with figures for fiscal 1989. Note discrepancies in dates of fiscal year. Onward reports a fiscal year from 1 March through 28 February. Renown follows the calendar year. Gunze reported data in 1981 according to a fiscal year beginning 1 December and ending 30 November. I used Gunze figures for the period, 1 December 1979–30 November 1980.

[a]Ratios in parenthesis represent comparison with a base of fiscal year 1980. Employment at Onward, for instance, declined 18 percent from the base year total of 4,013 workers in 1980, to 3,258 workers in 1990. Assets increased in the same period by 110 percent, rising from a level of $791 million in 1980 to $1.7 billion in 1990.

Table 9. Selected wholesalers: Performance, 1990
(in billions of U.S. dollars)

Firm	Sales	Profits
Onward Kashiyama	1.4 (+.38)[a]	0.065 (+.53)
Renown	1.6 (+.12)	0.022 (–.38)
Gunze	1.2 (+.19)	0.049 (+.53)
Wacoal	0.8 (+.24)	0.043 (–.17)

Sources: Please see the source note to table 5. Discrepancies in dates of fiscal year are explained in the source note to table 8.

[a]Ratios in parenthesis represent comparison with a base of fiscal year 1980. Sales at Onward, for instance, increased 38 percent between 1980 and 1990, and profits increased 53 percent.

60 percent of total sales at Toray, Toyobo, and Nisshinbo, but apparel represents close to 100 percent of sales at the fashion houses.

Unlike the pattern of ownership by a variety of banks and insurance companies at the spinners, the trading houses and the fashion houses demonstrate strong interfirm connections, including *keiretsu* ("family of firms") links. Onward is the exception; an educational foundation holds about 6 percent of its shares, a familiar list of assorted banks and insurance companies hold most of the rest. Renown retains keiretsu ties with the Sumitomo Bank, Life Insurance, and Trust companies among its top five shareholders.[1] Mitsubishi Trust and Sumitomo Trust are the leading shareholders in Gunze with 5.3 percent and 4.5 percent of its shares. Mitsubishi has also retained a place on the board, together with Dai-ichi Kangygo Bank, also a major shareholder.[2] One finds keiretsu links also at Wacoal, where Mitsubishi Trust and the Mitsubishi Bank own 12 percent of the shares. Deeply embedded in families of firms, including banks and commercial and manufacturing ventures, the pattern of ownership at the fashion houses provides a base and precedent for interfirm contacts.

The anomaly of finance at the fashion houses suggests a further contrast with mills and traders, and again highlights market exposure. The

1. Toyo Keizai, '92 *Kigyō keiretsu sōran* (An overview of enterprise keiretsu, 1992) (Tokyo: Toyo Keizai, 1992), 452.
2. Toray owned 1.1 percent of the shares in Gunze in 1990.

ratio of debt to assets at the fashion houses falls below the same ratio at the mills. Big mills can obtain credit on the collateral of their plant or real estate holdings, but little stands between the fashion firms and markets that might provide collateral to support loans. Apparel companies have neither the extensive real estate holdings nor a comparable base of plant and equipment to sustain more extensive debt burdens. And if fashion coordination and market coordination suggest continuities between apparel houses and trading houses, differences in function and scale explain radically different levels of debt. The ratios of debt to total assets for fiscal 1989 at Onward of 11 percent, Renown of 17 percent, Gunze of 24 percent, and Wacoal of only 9 percent fall far below the same ratios at the highly leveraged trading houses. Major trading houses finance the production process among smaller weavers, knitters, and wholesalers. Moreover, Itohchu enjoyed sales of $146 billion in 1990, which generated plenty of business in letters of credit and currency exchange for its main banks. Major fashion firms neither compete with the big trading houses in providing extensive credit to coordinate the production process, nor do they bring such brisk business to those banks. Renown enjoyed sales of only $1.6 billion in 1990, generating far fewer daily transactions for its main bank.

Turning from firm to function, we find fashion rather than production or marketing defining the giant apparel firms. The segmented structure of the textile industry leaves an opening for the special skills of the fashion firms. Mills do not often turn out finished goods such as apparel, and trading companies deal in larger volumes of fabric and yarn, rather than in designing commodity products for the fashion market. Coordination of the production, procurement, and marketing of apparel falls to Onward and its peers. Neither maker nor merchant, the apparel giants must coordinate the production and sale of fashions across a complex of makers and merchants, including larger and smaller spinners, finishers, and sewers, as well as department stores and larger boutiques. Onward and Renown are responsible for a diverse line of specialty, high-fashion, and mass-produced items. Onward describes itself as a garment wholesaler, but also advertises its limited garment-making operation as "under license from a number of the world's top fashion houses."[3]

3. Nihon Sen'i Kyōkai, ed., *Sen'i nenkan 1990* (Textile yearbook, 1990) (Tokyo: Nihon Sen'i Shimbunsha, 1990), 585; Onward Kashiyama, *Onward Kashiyama Co., Ltd. Annual Report 1991* (Tokyo: Onward Kashiyama, 1991).

Procurement helps distinguish the fashion houses from mill and trader. Suppliers for Onward include Itohchu, Marubeni, and other large trading houses.[4] Renown produces about 26 percent of the products it sells, but must subcontract with outside firms for the rest.[5] Itohchu, Tomen, and others bring the firm both yarn and finished goods from the major mills, often financing the sales of yarn from the mills to more specialized weavers.[6] A firm like Renown in turn subcontracts directly with the weavers who produce specialized goods with fancy designs such as jacquar or prints. As a fashion leader, the company concentrates on creating and marketing fashions, and only secondarily on sewing garments to ensure quality and reliable supply. In contrast, the mills concentrate on the manufacture of fiber and yarn, and only secondarily on weaving yarn into fabric to expand their product line.

Mogul and maverick spinners and fiber-makers give high priority to know-how and technology in the production process, whereas traders look to expertise in marketing and financing. Fashion firms compete in the organization of the production process, as evident in the emphasis at Renown and Onward on expertise in market trends, product design, and production. Renown devotes its own manufacturing capacity to products where quality and reliability are most important and must procure mass-produced, basic designs from the major spinners and larger weavers. For instance, the company now manufactures better than 40 percent of the women's clothing that it sells, but a lower percentage of the men's and children's clothing that it sells. Products make a difference in coordination of fashions, and makers of undergarments face a different challenge in production. Because Wacoal and Gunze limit production and sales mainly to men and women's underwear, they can devote their resources to producing large volumes of high-quality goods.

A narrower and less fashion-sensitive product profile at Gunze and Wacoal makes extensive production more feasible than at Onward and Renown with their more diverse fashion lines. Wacoal prides itself on being the "world's largest manufacturer and marketer of intimate

4. Dodwell Marketing Consultants, *Retail Distribution in Japan* (Tokyo: Dodwell Marketing, 1988), 401.

5. Renown, *Renown Incorporated, Annual Report 1990* (Tokyo: Renown, 1991), 4.

6. Nihon Sen'i Kyōkai, *Sen'i nenkan 1990*, 646.

apparel for women."[7] The firm itself operates only one garment-making plant, but retains eight wholly owned sewing subsidiaries across Japan. Gunze also prides itself as an in-house operation. The firm purchases combed cotton yarn of 30, 40, and 50 counts, then does its own weaving, knitting, and dyeing at the firm's ten wholly owned production subsidiaries across Japan. Indeed, Gunze has become notorious for its relentless demand of ever more consistent qualities of raw-cotton supplies. The firm itself will purchase the raw cotton from leading trading houses, and after careful inspection, contract the cotton out for spinning at mogul and maverick mills. Gunze does about half its knitting with cotton and half with synthetics purchased from the mogul synthetic-fiber makers. Toray, Asahi Chemical, and Teijin provide synthetic fiber for the firm, and Kondo Spinning, Nisshinbo, Toyobo, Nittobo, and Kurabo supply much of the cotton yarn.[8] One wonders why Gunze does not turn to cheaper suppliers of yarn and fiber abroad. Gunze executives argued that with a large volume of purchases from the same local mill, they were able to get a competitive price. Of equal importance, they could rely on the high quality of the local product. Again factors such as scale, reliability, and long-term relationships help keep the mills competitive in this market, despite lower prices of imported fiber and yarn. Unlike Renown and Onward, Gunze and Wacoal are dedicated manufacturers of mass-produced textile products. Like Onward and Renown, both Gunze and Wacoal devote extensive resources to the task of wholesaling their products.

If production and procurement define one role of the fashion firms, marketing is quite another. Fashion firms must somehow find a way to create and produce the latest fashions while holding down costs and retaining a reliable clientele of large-volume buyers. Thus fashion firms have developed cohesive and sophisticated marketing networks to bring their textile products directly to retailers. Mogul and maverick spinners and fabric makers sell mainly to trading houses or first-level wholesalers, and trading houses usually deal in large volumes with a variety of wholesalers, but fashion firms sell directly to the

7. Wacoal, *Wacoal Corp. Corporate Profile* (Kyoto: Wacoal, 1991), 2; Okurashō (Finance Ministry), *Yūka shōken hōkokusho sōran—Kabushiki Kaisha Wacoru* (A compendium of financial reports: Wacoal Corporation, Ltd.) (Tokyo: Okurashō, 1991), 15, 17.
8. Nihon Sen'i Kyōkai, *Sen'i nenkan 1990*, 360.

retailers. Onward counts department stores across the country among its customers, as well as boutiques and general retailers. Isetan, Seibu Department Stores, Takashimaya Company, Tokyu Department Stores, Sogo, Mitsukoshi, Daimaru, and Matsuya all appear among Onward's customers.[9] Onward also has direct contracts for uniforms with government offices and private companies.[10] Renown markets 47 percent of its products through department stores, 40 percent through specialty shops, and 8 percent through chain stores and supermarkets.[11] They insist that "bypassing intermediate distributors not only decreases costs, it also speeds the marketing process, which gives Renown an edge in reacting to fickle fashion trends."[12]

Fashion defines at best a variety of interests at the apparel houses where profits, growth, and survival depend on rapid response to changing consumer tastes. The creation of fashions bridges production, procurement, and marketing. For instance, four planning and production cycles, corresponding to seasonal fashions, continue simultaneously at Renown. Planning for the fall market begins in the previous September with a review of the performance of Renown's fall fashions for the past year.[13] Soon specialists in materials, color, design, fabrication, and marketing begin to share ideas for the coming season and invite submissions from the larger spinning firms, trading houses, and especially subcontractors. Renown will produce a sample jacket or blouse by January in one representative color and pass it on to the sales personnel for discussions with key buyers. After initial testing of the market and further discussion between sales and merchandising personnel, a preliminary plan for production is worked out in February. Then the real work begins.

Company officials spoke of the difficulty of coordinating market analysts, producers, salespeople, and a network of subcontractors. One exasperated executive struggling to corral subcontractors complained: "I have responsibility for oversight of design, measurements, colors, and schedule all at the same time." Renown factories and subcontractors then have only six weeks or so to produce model items in a variety of colors for an exhibition in March. Now the buyers choose

9. Dodwell, *Retail Distribution in Japan*, 401.
10. Nihon Sen'i Kyōkai, *Sen'i nenkan 1990*, 585.
11. *Renown Incorporated Annual Report 1990*, 4.
12. Ibid., 4.
13. The process is outlined in *Information '92 Renown*.

among the new fashions and colors and make their orders. On the basis of orders received, Renown then formulates a final production plan in April with deliveries four months later. Any breakdown in the year-long process will delay deliveries, and if deliveries do not meet deadlines for seasonal sales at the retailers, buyers for the large-volume retailers will find other suppliers. Consumer demand for product reliability and on-time delivery help maintain long-term relationships with the department store and chain-store purchasing agents downstream, and large-scale, reliable mills upstream.

Less extensive fashion changes in underwear distinguish the marketing tasks of Wacoal and Gunze from the challenge of fashion leaders such as Onward and Renown. Firms like Wacoal and Gunze produce a consistently high-quality product with little seasonal variation for volume sales and then specialize in smaller-lot items for specific markets. Wacoal markets through department stores, chain stores, and specialty stores and even offers special training for saleswomen at major department stores.[14] Gunze has set itself apart from Renown and Onward with comprehensive integration of production, wholesaling, and even retailing. The firm produces a limited line of knitted goods, but it carefully manages and controls the knitting process from fabrication to distribution entirely within its own network of plants, distribution centers, and sales outlets. Gunze markets about half of its production through its own outlets and about half through wholesaling. Chain stores such as Daiei, Seiyu, and Itoya stand among their best customers, rather than department stores, which prefer higher-count cottons. The firm also markets goods through major trading houses such as Mitsui, Mitsubishi, and Marubeni.[15]

Their marketing tasks distinguish the fashion firms from the mills; their production distinguishes them from the trading houses. Manufacture and marketing "under one roof" demands quite a range of expertise at the fashion houses. Lacking extensive intrafirm plant facilities, the fashion houses must integrate planning, design, production, and delivery across a network of subcontractors. Their expertise is not only in markets or "fashions," but also in organiza-

14. *Wacoal Corp. Corporate Profile,* 8.
15. Nihon Sen'i Kyōkai, *Sen'i nenkan 1990,* 359; Gunze, *Gunze Today* (Osaka: Gunze Ltd., 1991).

tion of a production network for specified products in a very short period of time. Promoting sales of their products to retailers demands a quite different kind of marketing expertise and structure. Fashion exhibitions bring together regular clients for Onward and Renown, particularly buyers for the big department stores, whereas Wacoal and Gunze rely more on their own distribution and wholesaling systems. The fashion firms bring a precious expertise to the industry in merchandising such as prediction of market trends, product design and specification for manufacture, and coordination of the garment-manufacturing process.

"Coordination of fashion" best conveys the multiple functions of the fashion houses, and their particular expertise in bridging interests from apparel maker to retailer. The profile given earlier suggests a continuity of scale, market leadership, commitment to textiles, and market exposure among the four firms. Commitment to textiles distinguishes Onward and others from the mills and trading houses and leaves the fashion giants with a very specialized and scarce expertise. Competition with lower-cost foreign competitors has forced the fashion houses to curtail their sewing operations radically, leaving them dependent on contracting rather than in-house manufacture. Procurement demands quite different skills of coordination than production, since fashion houses live or die by how rapidly they adjust to market demand and how reliably they deliver. With little collateral for loans and less opportunity for long-term contracts because of changing product lines, fashion houses must coordinate and adjust simultaneously.

Adjustment for the moguls has been encouraged in part by insulation from market fluctuations. More direct exposure to consumer markets distinguishes the fashion firms. Mills deal mainly with trading houses and larger wholesalers, who sell in turn to weavers, knitters, and garment makers, who will always need yarn and synthetic fibers. Downstream fashion firms find less security in the demand for their market mediations, for although retailers and their customers will probably always want the latest fashions, they may well find procurement channels more efficient than the apparel giants. Fashion firms must be able to sell their own products in fickle markets to department stores, chain stores, and specialty stores. Apparel firms such as Renown and Onward must constantly adjust to market demand by renegotiating products and production schedules with their contractors. Embedded ties with networks of spinners, weavers, finishers, and sewers help sustain the fashion houses, but also demand adjustment of products and renegotiation of common goals.

Commitment and exposure suggest a further feature of structure and interest at the fashion firms, where continuity must be more closely balanced with competition. Priorities of scale, reliability, and variety of products help sustain ties between the major mills and the larger trading houses. One can point to a similar coincidence of priorities joining Onward and its peer fashion houses to the major mills, fostering continuity, survival, and even prosperity. But fashion giants must compete with trading houses that can broker or sell commodity products of the mills across multiple markets within and beyond Japan. Fashion houses often contract for more specialized yarns or fabrics, and must then broker such products along a narrower line of large-scale retailers within and beyond Japan. Continuity of scale and reliability between fashion house and mill depends on the adaptation of mill to market, rather than solely on consistent production of large lots of similar products for stable markets. Commitment and coordination, market exposure and interfirm contracting, and continuity and competition help distinguish the structure and interests of fashion houses in Japan's textile transition.

Adjustment

Fashion firms offer a foil for the mills. Moguls and mavericks share a common task as upstream producers of textiles, but the four leading apparel firms offer four different blends of garment making and wholesaling. These firms find continuity in fashion rather than in any single function. Mills compete for markets in yarns, fibers, and fabrics with similar qualities of cotton and often similar blends of natural and synthetic fibers. Fashion giants focus on distinct, specific segments of the apparel industry, with some overlap between huge firms such as Onward and Renown. Mills remain leaders in a relatively concentrated production sector, but fashion houses compete with a wide range of larger and smaller wholesalers and garment makers. Finally, moguls and mavericks command the heights of the production process, churning out yarn, fiber, and fabric that downstream firms need for weaving and knitting, finishing, and garment making. In contrast, fashion houses stand at the end of a long production line, making and selling seasonal fashions directly to retailers. Market exposure and commitment, and a specialized expertise in the coordination of fashion, distinguish these merchants from the mills.

Despite their differences, merchant and mill share a similar vision of adjustment and face a similar challenge from lower-cost producers abroad. But mills can add to their commodity product line with specialized yarns, use their capital and plant resources to diversify out of textiles, or move production offshore. Fashion houses have reached a limit on the balance of commodity and specialized product lines and lack a comparable base of resources for diversification. Moving to offshore markets and production appears the most viable path of adjustment for the fashion houses dedicated to the industry, but the transition has been slow. Renown and Gunze prefer in-house manufacture to maintain their reputation as producers of high-quality and reliable products. Gunze maintains enough capacity within Japan to mass-produce large volumes of men's underwear of high-quality combed cotton, and Renown can manufacture women's clothes to meet the higher standards of the women's fashion market in Japan. Yet Renown has reduced its manufacturing capacity across the past decade, as has Gunze. Both firms face a decline in local production capacity caused by the same difficulties with wages and labor shortages plaguing the mills. Officials at Renown complain of the difficulty in just finding workers for a garment-making operation. I raised the same issue with Gunze executives who also pointed to a labor shortage. Capital-intensive machinery in the knitting and dyeing operations permits a reduction in the labor force, but not in sewing, which remains largely labor-intensive. They spoke of an aging female population of sewing-machine operators, working on their feet most of the day, handling up to three machines at once. Since the firms today can no longer attract younger workers to replace the operators, both Renown and Gunze have moved production operations abroad to cope with the labor shortage and rising local labor costs.

Moguls and mavericks have invested abroad to continue and expand their traditional role as synthetic-fiber producers or spinners. Traders have invested in offshore production to supply their export networks or penetrate local commercial networks in foreign countries, in line with their familiar global trade strategies. Since fashion firms cannot so easily transfer their expertise abroad, they have traditionally invested in only a limited phase of the production process. Onward and Renown operate sewing plants, and Wacoal and especially Gunze are active in knitting and dyeing, but definitely not in spinning. Unlike the traders, the fashion firms have traditionally enjoyed neither the expertise nor the capital to finance production systems across national borders. Fashion houses often straddle the functions of both production and sales abroad, just as they focus on both manufacture and marketing at home. Coordinating fashion across national borders presents the apparel firms with special problems of adjustment and interest mediation in offshore investment.

Coordination of fashion within Japan includes marketing, procurement, and production, and the challenge of adjustment has nudged the firms to extend these functions beyond Japan. For example, Onward, Renown, Wacoal, and Gunze all market their goods abroad, all subcontract production abroad, and all but Onward have joint investments for production abroad. The key to all three strategies is reliable and cost-effective production outside Japan. Onward operates a marketing network of "boutiques and shops in Europe, North America and Asia" and has recently joined Itohchu in a sewing operation in Talien, China.[16] Renown has invested in both marketing and production abroad, and Wacoal has used its advantage in manufacturing expertise and limited product line to invest in production and marketing in Asia. Marketing goods abroad without manufacture is a costly process, for transportation costs necessary for extensive sales abroad in the absence of offshore production facilities will quickly erode the profitability of overseas sales networks.

Fashion firms might adjust to lower-cost competitors abroad with procurement from offshore suppliers, but factors such as product reliability and cost in the production process determine the success of both procurement and production abroad. A fashion firm can subcontract for either finished goods or basic materials, but uncertain quality, reliability, and variety discourage importing the former. Officials at Renown observed that only a very small percentage of their imports were finished goods ready for sale. Instead, the firm contracts for fabric and yarn from a network of spinners and weavers

---

16. *Onward Kashiyama Co., Ltd., Annual Report 1991*, cover page; Toyo Keizai, *Japan Company Handbook, First Section, Spring 1992* (Tokyo: Toyo Keizai, 1992), 896.

abroad, and then sews the apparel in their own factories or through their networks of local subcontractors. Materials from their offshore contractors further complicate coordination of the production process. For example, officials at Renown complained that yarns and fabrics from abroad do not hold dyes as well as locally produced textiles do.

The dilemma of fashion coordination across borders is clear. Fashion houses sell their products on the basis of color, fabric quality, and design, and Japanese consumers will not tolerate inconsistencies in fabric or yarn. At the same time the higher cost of domestic fabric and yarn forces them to seek out materials from abroad. The problem confronting the fashion firms is effective transfer of their expertise in design and organization of production to foreign production sites. Trading houses hope to develop subcontracting networks with Indonesian garment makers, but must rely on market-sensitive buyers in Japan to specify the design, materials, color, and sizes and to take some part in overseeing the production process. Fashion firms such as Onward and Renown have long experience in fashion, design, and oversight of production processes and would appear ideally suited for coordinating garment making overseas. Thus far, fashion houses have survived foreign competition in Japanese markets with yarn and fabric imports for use in their lower-range products, but rising costs of domestic sewing operations may soon force a shift of garment operations abroad, prompting new challenges in the coordination of design and fabrication.

Unlike subcontracting, integrated production abroad at affiliates or subsidiaries permits closer control of manufacture and design. About fifteen years ago Renown bought production facilities in South Korea and Taiwan aimed at local market penetration. Rising labor costs in both countries have led to more capital-intensive production of higher-quality goods at both subsidiaries and a reorientation to the Japanese market. Subsidiaries in Hong Kong and Singapore do some sewing, but serve mostly as sales outlets for Renown. Wacoal and Gunze have a product advantage in foreign manufacture over their wholesaling rivals, for their underwear lines are less diversified and thus less sensitive to changes in fashion. Both firms have begun investing in production abroad aimed at both local and foreign markets. Wacoal has developed marketing networks in Seoul, Taiwan, Hong Kong, Singapore, Bangkok, and now Beijing, as well as in the United

States. Wacoal also manufactures its goods in China, Taiwan, Thailand, Barbados, and Puerto Rico.[17]

The history of efforts abroad at Gunze across the past two decades perhaps best exemplifies the challenge of fashion coordination across borders. Gunze has a reputation for demanding high standards, whether in supplies of raw cotton or in fabric and yarn. Local inter-firm networks with major mills and trading houses can accommodate this preoccupation with quality and control, but can Gunze find comparable networks abroad? Offshore production in the developing economies of Asia is not an attractive alternative. Nonetheless, Gunze established a joint venture in Korea as early as 1971 for the production of underwear and another, more recently, in Thailand, also for underwear.[18] Apart from apparel manufacture, Gunze has also developed a supply network with joint ventures in Hong Kong for production of sewing thread from 1972 and for industrial-use thread in Indonesia from 1991.[19] Mills go abroad with trading houses, but fashion houses often go alone and find partners abroad. Gunze officially provides roughly half the equity for these ventures, and local partners the rest.

Despite these foreign ventures, Gunze officials estimate that goods produced abroad accounted for only 10 percent of total sales through 1991. There were also distinctions. The firm marks the foreign-made products with a green label to distinguish them from locally produced goods with a brown label. Nonetheless, the cost of labor, yarn, and fiber in Japan is forcing even Gunze to consider moving production abroad or perhaps diversifying. The firm has made only limited headway with investments apart from apparel, with nontextiles accounting for a mere 18 percent of its total sales in 1991. It remains to be seen whether Gunze can now find the capital to invest in expanded pro-

17. Toyo Keizai, *Kaisha betsu kaigai shinshutsu kigyō 1990/1992* (Enterprises investing abroad: Investments listed by firm) (Tokyo: Toyo Keizai, 1992), 143–44.

18. Toyo Keizai, *Kaisha betsu kaigai shinshutsu kigyō 1991/1992* (Enterprises investing abroad: Investments listed by firm) (Tokyo: Toyo Keizai, 1992), 122–23. The firm in Korea is titled "Chonbang Gunze," in Thailand "Thai Gunze," in Hong Kong "Gunzetal Limited," and in Indonesia "P. T. Gunze Indonesia." The enterprise in Korea now includes six hundred employees.

19. The firm also maintains technological cooperation agreements with textile firms in Taiwan, China, Malaysia, and Thailand. (Under a technological cooperation agreement the Japanese firm receives royalties for the training it provides and for the use of its technology.)

duction abroad or adjust to constraints on quality control abroad and still satisfy the demanding consumer market in undergarments it has helped create. The firm has moved aggressively to expand sewing operations abroad in the past five years, while still controlling the knitting and dyeing processes at home. But transportation costs of exporting knitted goods for sewing, and then bringing the garments back to Japan for sale have made this alternative less profitable. Officials at Gunze predict they will be doing more integrated production abroad in the future.

Investments in offshore marketing and production shed light on the comparative strengths and market strategies of mills, traders, and fashion houses. Moguls can bring their expertise in machinery and production processes to a joint investment abroad, usually in tandem with the financing and marketing expertise of a major trading house. Both mill and trader contribute capital and expertise, and both profit from the textile production of the firm abroad. Fashion firms cannot easily export their talent and experience. They are quite adept in predicting fashions for the Japanese market, but less competent in surveying and predicting tastes in the developing economies of Asia.[20] Fashion firms also cannot easily export their expertise in designing and overseeing the production process of garment making. Officials at both Renown and Gunze complained of quality control over production abroad, a hallmark of their success in production within Japan. An industry-wide vision of adjustment for the textile industry encouraged offshore investment, but firms in upstream production sectors can benefit more easily from such a change than the downstream fashion firms can.

Adjustment at the apparel giants again suggests market exposure, market commitment, and fashion coordination. Apparel firms often venture offshore alone, without the expertise or financial clout of the trading houses, and have generally limited their investment abroad to apparel, in line with their distinctive expertise. Nonetheless, these houses of fashion apparel have extended their role of coordination to new international horizons with cross-border investments. The review above of moguls and mavericks alerted us to a pattern of direction and divergence within the sector of spinning. We find now a parallel

20. Wacoal is an exception with their investment in marketing women's foundation garments for markets in northeast and southeast Asia.

of direction and divergence not within, but rather across sectors of mill and merchant in the textile transition. Both fashion-apparel giants and the major mills have joined in shaping and following a common compass of change that includes specialization, diversification, and offshore production. The latter effort abroad suggests a continuity across mill and merchant, but it also reveals differences in their respective expertise and scope of investment. Constructing and operating a spinning mill or synthetics plant abroad demands a huge, long-term investment, whereas a garment factory demands far less capital and commitment.

Apart from adjustment, the effort of apparel firms abroad indicates new efforts to shape interfirm interests across borders. Mills and traders go abroad in tandem in line with earlier precedents of marketing ties. Itohchu formerly brought Toyobo products to Indonesia, and now joins Toyobo in establishing P. T. Unilon for local production. Mill and trader develop new bases for shared goals on the basis of earlier experience. Gunze, Renown, Wacoal, and other apparel giants more often move offshore without a Japanese partner and must develop new bases for shared interests with foreign suppliers or production partners. Both dedicated to the industry and exposed to intense market competition, the apparel firms must find new organizational ties to enhance their role as coordinators of fashion beyond as well as within Japan. Only with effective adaptation to this new role of interest mediation across borders will the fashion houses continue to sustain the moguls and mavericks and bring mill to market.

Fashion firms extend the comparison of coordination, interest accommodation, and corporatism in the textile transition. The function of coordination or mediation reflects both distinctive structures and distinctive interests. "Authoritative coordination" by the state joins public interest in employment and regional development with industry interest in economic performance. The expertise, resources, and legal authority of the Ministry of International Trade and Industry ensure a public interest in the tripartite negotiation of adjustment with capital and labor. "Market coordination" of the GTC bridges the expertise of specialized producers in spinning, weaving, and production of synthetic fiber with the marketing capacities of first-level wholesalers. The market contacts and the financial resources of the trading houses ensure long-term market opportunities for large-scale producers of fiber, yarn, and fabric. "Fashion coordination" at Onward and others joins the expertise of apparel producers with that of retailers, serving the interests of both producers and fashion consumers. Organization in planning, production, and sales gives the fashion houses the necessary structural resources to bridge interests and coordinate the fashion process from mill to apparel maker to retailer effectively.

Accommodation of interests between mills and fashion houses recalls familiar priorities of reliability, variety, and volume. Larger mills can deliver yarn, fiber, and sometimes fabric in the quantities and time frames necessary to meet the demands of the fashion houses. Fashion houses or other brokers must then contract out for dyeing and finishing, as well as sewing, before the products are ready for delivery to the retailer. Continuities of scale and productive capacity reinforce long-term ties between the major fashion houses and major mills, helping to sustain the survival of the moguls without discouraging emergence of the mavericks. Vigorous growth among the fashion firms in the downstream sector of the textile industry in recent decades, together with long-term relationships with the mills for reliable, high-quality, large-volume supplies helped moderate decline among the upstream producers. Yet greater sensitivity to changing tastes in apparel at the fashion houses have brought new demands to their suppliers. Fashion houses can maintain large-scale contracts with mills adapting their yarns and fibers to the demands of the market, but find less continuity with mills producing solely the lower quality, commodity products. The maverick Kondo Spinning in Nagoya, for instance, can sell its higher-quality combed cotton to Gunze, but not its lower-quality carded cotton.

Marketing, procurement, and production abroad has added a new dimension to fashion coordination at the four major firms. Interest accommodation has stretched the fashion houses beyond their traditional links with Japanese mills to networks abroad. Fashion houses must now find common ground with a network of suppliers and retailers not only within but beyond Japan. Just as the mills and traders must accommodate the interests of local partners at P. T. Unilon in Indonesia, or Luckytex in Thailand, fashion houses in the textile transition must coordinate multiple partners in the apparel production line abroad and then negotiate contracts with retailers in Japan and elsewhere in Asia. Extending intrafirm goals among three major partners at a single mill is one thing, but extending intrafirm goals among multiple partners across borders in the joint task of design and manufacture of changing commodity fashions is quite another.

Furthermore, the chronicle of change among apparel firms offers additional insight into corporatist strategies of adjustment. Accommodation of interests as fashion coordinators permits a market access that ensures continuity of supplies from the upstream producers. Continuity of markets, in turn, absorbs much of the production of the leading mills, whether mogul or maverick, helping them to weather the transition out of textile production, or to find a bigger market for their expanded production. Here the fashion houses support both direction and divergence in the flexible corporatist strategies of adjustment apparent among the mills. Fashion houses also appear as foils for the mills in interests and adjustment patterns, in market exposure, industrial commitment, and expertise. More important, there is a contrast between interfirm cooperation in corporatist strategies of change upstream and the absence of cooperative strategies of change among the fashion houses downstream. The loosely affiliated fashion houses stand in sharp contrast to the structured interests of the mills. Continuity of interests at the mills in yarn and fiber production, their shared concern for maintaining or expanding market share here and abroad, and their joint priority of maintaining dominance in spinning sustains common bargaining positions with labor and state and collective action in the adjustment process. A diversity among fashion firms in function and markets erodes the consensus necessary for collective actions to moderate changes in the market. The segmented structure of the apparel sector likewise does not facilitate common action.

The lack of clear identification either as wholesalers or as garment makers compounds problems of common action within an employers' association. Relatively weak bases for cooperative action among the fashion firms are evident in their ties with mediating organizations. Executives at Renown and at Gunze could not suggest any association of their own comparable to the JSA or the JCFA. Labor relations among the fashion firms also provide no continuities. Employee unions at Gunze and Renown affiliate with different sections of Zensen, and workers at Onward and Wacoal maintain affiliations with other federations. State offices such as the Ministry of International Trade and Industry can look to no common sectoral identity among the fashion firms for maintaining regular policy consultations and information exchange. What might explain the relative absence of corporatist strategies of change among the downstream apparel firms? We could cite lack of historical precedent for a common structuring of interests, or of a clear sectoral function shared by all the firms. Concentration or dispersion within a sector also affects common action. We can cite here a multiplicity of smaller competing apparel firms, and also market exposure. The fluidity of downstream retail markets also discourages the long-term, common structuring of interests evident among the mills. The evidence suggests conditions supporting societal corporatism in declining industries, such as a clear sectoral identification, relative concentration and comprehensive representation within a sector, and commitment to a strong industry association.

9 Collective Action

One might analyze the ties within and across firms, sectors, and industries that insulate adjusting mills from the vagaries of the market, but the reality remains an embedded pattern of institutions and routinized practices that channel conflicts and moderate unruly market dynamics. Moguls pursued a common compass of change, though with variations in emphasis on product specialization, diversification, and offshore investment. Mavericks opted rather to reinvest in spinning, but also pursued distinctive strategies of offshore investment and limited diversification into real estate. The independence of the Japanese firm, particularly of the privately held mavericks and the financially secure dissenters like Nisshinbo, allowed new market commitments in spinning, at the same time most of the moguls sought refuge from that same market. The cooperation of labor supported such independence. State encouraged a common compass, but had few resources to encourage a reduced commitment to spinning and no legal power to force reductions.

Analysis of sectoral dynamics brings the issue of direction and divergence into sharper focus. The key to collective action here is not what individual firms did to specialize, diversify, or move offshore. Rather it is the joint commitment to market insulation to ensure steady profits from their commodity production of yarn, fibers, and fabric. Capacity-reduction programs and production cartels provided insulation and helped bring balance to supply and demand, but also drew the ire of dissenting moguls and mavericks. One wonders why divergence of opponents did not undo the convergence of the supporters of a common direction of adjustment. "Industrial citizenship" provides one answer and adds further insight into the roots of cooperation in Japan, for opponents carefully maintained their roles as productive, contributing members of the industry associations. They followed legal procedures in expanding capacity as the moguls reduced capacity and dutifully reported their expansion of capacity to maintain a public record of dissent.

Intersectoral links between mills and merchant extend the question of direction and divergence to market ties beyond individual firm and association. The key here is competitive access of mogul and maverick alike to the giant fashion houses. Apparel firms cared little about direction and divergence at the mills, but cared greatly about reliable supplies of yarn, fiber, and fabric. The fashion firm needed a large-volume supplier, reliable deliveries based on long-term relations, and consistent product quality. Priorities of scale, reliability, and variety linked mill and merchant, providing opportunities but not necessarily contracts. Individual spinners had to choose to pursue ties with the fashion firms, offering Gunze the kind of high-quality combed cotton necessary for its product line, or serving the more diverse product lines of fashion giants like Renown and Onward. A reliable source of revenue was one benefit for moguls and mavericks burdened with new investments to ensure effective adjustment in a declining industry. Mills could diversify into nontextile production in part because wholesalers continued to purchase their textile products, providing income necessary for investments in diversification or offshore production. Mills could specialize their products because wholesalers and retailers had created a market for those textiles.

Ties with banks and traders further sustained a common direction of adjustment among the moguls despite divergence by the mavericks. Prospects of decline in the industry may well have driven banks away from the industry without the pattern of deeply rooted ties between bank and manufacturer. Banks served as the major owners and major customers for the banking business of the moguls. Close ties prevented capital flight, but also helped persuade the banks to provide capital necessary for adjustment. Apart from funding, there may also have been pressure from the banks for adjustment programs, since the powerful main banks of the moguls would probably not have tolerated retrenchment from the common compass of change. Mills need incentives to change. Moguls with deep traditions as spinning firms may well have opted to stay with an industry they knew, rather than diversify into unfamiliar areas of nontextile production or move into textile production abroad in areas they did not know. But the banks and insurance firms were committed to profitable growth at the moguls, not to the spinning industry per se. The banks could not bring such pressure to bear on the more secure Nisshinbo, or on the privately held Tsuzuki or Kondo Spinning.

Ties with trading houses appear critical to the survival of the mills. Cooperation in imports, exports, and offshore investment signal deeply rooted, long-term relationships between mills and these merchants. Like apparel firms, traders appeared less concerned with adjustment paths at the mills than with performance. Yet the long-term, embedded network of ties between mill and trading house sustained the moguls and still permitted the emergence of mavericks. Trading houses offered consistent market opportunities for the moguls and mavericks within and beyond Japan during the years of adjustment. Itohchu and others maintained a network for reliable supply of raw cotton at competitive prices and for offshore investment in production, which supported both continued production at home and the transfer of production expertise abroad. Individual firms used the traders' network to their advantage in specializing cotton products, or in marketing commodity yarns and fabrics.

Banks and merchants in trade and fashion helped sustain survival and hierarchy among the moguls without excluding the rise of mavericks. Divergence among a minority did not disrupt the common direction of adjustment for the majority, and citizenship at the association level provided a recognized channel for divergence. The independence of the firm itself permitted individual choices for market insulation or exposure during the years of decline. Ties with merchants and banks provided a pattern of flexibility that helped define the cooperative umbrella of adjustment. State and association could accommodate divergence within this cooperative framework. Mavericks and a few dissenting moguls found considerable room for market choices within this umbrella, but the majority of moguls found security from market disruptions during the vulnerable years of specialization, diversification, and offshore investment.

Moguls and mavericks have provided our central insight into the adjustment process. What is most significant about the maverick dynamic in interfirm relations is its relationship to the coalition of capital directing the process of change at the firms. I have highlighted the flexibility and relative effectiveness of the employers' associations in working out a shared compass of change that could accommodate divergence. Other industries in Japan may find the precedent instructive. Challenges of adjustment for advanced industrialized economies can seldom be cast simply in terms of "sunrise" as against "sunset" industries. The more common problem will be capital-intensive production in formerly labor-intensive industries, where there remains potential for reducing fixed costs, creating new products, and developing high value-added markets. Various Japanese industries will confront the challenge of adjustment in the next decade, and some will have the opportunity to restructure production within the industry, rather than simply move out of the industry. Firms in highly concentrated industries may have less difficulty creating a consensus for change, but for industries that do not enjoy the concentration that permits more consistent strategies among a few dominant firms, the record of a porous yet effective solidarity at the textile associations may offer a constructive precedent for change.

Corporatism

The term "cooperative umbrella" may well suggest a structure or organization, rather than a process of mediation. Persistence of a common direction despite divergence may well turn attention to result, rather than to a chronicle of conflict and compromise. But a focus on the bridging of organized interests in structured relationships among state, capital, and labor draws us closer to procedure and process. Review of mediation offers an answer to our initial question about corporatism, and contrast with pluralism, or reciprocity. Patterns of reciprocal consent can be cited between state and capital, and to a lesser extent between capital and labor, in the textile transition. But can we say more of this curious agility at compromise? Samuels declined to link his findings with the wider debate between pluralism and corporatism and offered a sobering assessment of the two terms: "Their semantic entanglement has rendered them more confusing than enlightening."[1] I recognize the semantic ambiguities but still find heuristic utility in the concepts, particularly in the Japanese context of interest mediation, which often remains opaque to Western observers.

1. Richard J. Samuels, *The Business of the Japanese State* (Ithaca: Cornell University Press, 1987), 8, 292 n. 19.

An effort to sort through such terms can contribute to the debate about the shape of market societies in Asian capitalism.[2]

Neither neocorporatist nor patterned pluralist theories alone can explain the record of planned change across the breadth of the textile industry. Yet as I turn from political economy to a sectoral focus on textiles, I find the term "pluralism" misleading. One observes neither a multiplicity of competing interest groups nor even much mobility among the leading groups. Few would suggest solely a passive or reactive role among state bureaucracies concerned with this declining industry. Perhaps the most telling criticism of the pluralist argument lies not with state nor industry, but in their permeable boundaries. The pluralist thesis cannot account for interpenetration of public and private interest, or of interests across industry and firm, labor and management, merchant and mill.[3] Adjustment in textiles offers evidence of patterns of interest representation that might well fall within a patterned pluralist framework, but in addition to the objections cited above, the term "pluralism" too easily draws us back to unproductive assumptions in the study of Japanese society. Pluralism recalls features of the debate about capitalism in the United States, such as assumptions about competition as an unrestricted good, of ad hoc, issue-oriented interest associations, or of a government role limited to oversight in an arena of competition without state intervention.

The key issue remains effective patterns of cooperation that somehow promote a more competitive specialization among textile magnates. I have argued that corporatist strategies of adjustment sustained hierarchy among the moguls, but also permitted emergence of the mavericks. Corporatist strategies include structured, long-term relations between state and capital, with the interests of labor somehow accommodated to ensure cooperation and participation in adjustment efforts. Suggestions of corporatist forms of cooperation and pluralist forms of competition offer one resolution to the debate, but provide little insight into the collective structures and procedures within which firms compete for a market niche. That we find corporatist strategies

2. See Harmon Zeigler's *Pluralism, Corporatism, and Confucianism* (Philadelphia: Temple University Press, 1988).

3. For a summary of the neocorporatist critique of pluralist assumptions, see Noel O'Sullivan, "The Political Theory of Neo-Corporatism," in *The Corporate State: Corporatism and the State Tradition in Western Europe*, ed. Andrew Cox and Noel O'Sullivan (London: Edward Elgar, 1988), 5.

prominent in the decades of change among a disparate array of larger and medium-size spinners, and among firms with a long and proud history of independence from government direction, suggests the significance of this case of societal corporatism.

Corporatism denotes intermediation or bargaining, rather than simple open-ended interest competition. Corporatist ties in Japan's textile industry are hierarchical, institutionalized yet flexible, and societal. Hierarchy is apparent in the leverage of major firms in formation and enforcement of the industry-wide vision of change. The feature of "institution" underlines the importance of sustained organization apparent in the role of employers' associations, or of enterprise unions and labor federations in the corporatist strategies of information-exchange, policy formation, and enforcement. The term "institution" also turns attention to the underlying, rooted character of precedent and procedure. Flexibility refers immediately to the porous quality of solidarity among the often contentious spinners in the transition. But the wider significance of flexibility within societal corporatism lies in the art of compromise among capital, state and labor in Japan, which permits radically opposed market choices of exposure or insulation among firms within the same industry.

Corporatist theory helps explain patterns of collective action promoting competitive survival in a declining industry. Corporatism helps explain mediations that embed and extend special interests into more general, comprehensive interests within and across firms and industries. Corporatist theory draws attention to collective frameworks for sustaining the roles of capital, labor, and state within a changing political economy. The record of change in Japan's textile firm and union, state bureaucracy and Diet, local and foreign markets gives content to the thesis of a societal or neocorporatist strategy of change. Analysis of the Japanese experience contributes to a growing comparative literature on patterns of cooperation and competition within capitalist political economies. Since 1956 the state and the mills have brokered their structured interests to design and implement a collective program of adjustment in the upstream and midstream production sectors of the industry. Lest corporatism at mills imply corporatism across the industry, we looked to both market frame and corporatist foil in the garment industry. Reliable market access through long-term relations with leading fashion houses helped eased the adjustment for moguls and mavericks alike. The market framework or context supported reliable demand for yarns, fibers, and fabrics, among mills either diversifying out of the industry or reinvesting in the industry. A blend of direction and divergence in corporatist adjustment strategies upstream was made possible in part because of reliable market opportunities downstream. Yet looking beyond the interaction of mill and fashion house to inter-firm strategies of adjustment in the apparel sector, we found no evidence of corporatist links among fashion houses themselves, or between fashion firm and state in the same period. A base of similar interests and investment among mogul spinners and fiber makers, a tradition of strong employers' associations, and precedent help explain the corporatist alternative among upstream producers.

Two problems remain in efforts to clarify corporatist adjustment efforts in Japan. Labor has yet to secure an autonomous role within tripartite negotiations of adjustment. The combination of a strong federation such as the Zensen, and the recent emergence of the Rengo as a national center greatly enhance the potential for textile workers to assert a more independent role, but a clear case for corporatism in Japan must await a firm resolution to the problem of corporatism without labor. A second problem lies in the distance between public rhetoric and effective structure, for state, and even capital and labor take easy refuge in statements of social partnership, leading George Aurelia to conclude: "Corporatism as defined here has existed most clearly in postwar Japan as an ideal expounded at different times and in different forms by MITI, the Economic Planning Agency and by the Japan Committee for Economic Development. It has, however, been instituted only in limited form in the Japanese economy."[4] He would relegate corporatism to state rhetoric, but there is more than corporatist rhetoric in the textile transition. I document corporatist strategies of change across nearly four decades of adjustment in a spinning sector without extensive oligopoly. I specify corporatist structures such as the Textile Council, business associations, and labor federation, taking pains to distinguish between corporatism at mills and unmediated exposure to market dynamics among merchants.

Methodology offers some resolution to these problems in the study of Japanese corporatism. Analysis of corporatism demands attention to capital, state, and labor. The study of business/state ties in Japan alone

---

4. *The Comparative Study of Interest Groups in Japan: An Institutional Framework* (Canberra: Australian National University, Australia-Japan Research Centre), 66.

or of industrial relations alone limits our understanding of corporatism. Also needed is a comprehensive effort to specify microcorporatism, mesocorporatism, and macrocorporatism. Microcorporatism in Japan begins with the firm, the centerpiece of our study. We have specified mesocorporatism in the spinning sector, with attention to business association and labor federation. The Textile Council, Rengo, and state bureaucracies offered some insight into channels for macrocorporatism. Also, an emphasis on how interests are made complementary, rather than on complementary interests, helps maintain a focus on process rather than simply on results. A threefold analysis of corporatist strategies of change in Japan offers promise for resolving the dilemmas of labor and rhetoric.

What might the past tell us of the future? What remains of corporatist strategies with the fading of formal programs of adjustment in the industry? Corporatist ties have been both routinized and somewhat attenuated in continuing efforts to enhance the competitiveness of mogul and maverick firms, but nonetheless remain formidable structures for change and redirection in the industry. The exchange of information and mutual consultations on adjustment continue among capital, labor, and the state within the same government offices, business associations, and labor unions prominent in the earlier decades. Neither industry nor state would welcome corporatist strategies of cooperation in formal adjustment programs at present. State might find benefit in greater coordination, but lacks the rationale and resources for intervention. Mills might welcome state support, but not state oversight. The textile transition has brought changes in the comparative leverage of capital, state, and labor, which suggests corporatist structures also have changed. Nonetheless, a renewed sense of crisis, rededication to strategies of collective adjustment, and additional state resources for assistance programs and tax incentives could quickly revive the earlier corporatist coalitions.

Looking to changes in the state, the growing role of the Diet in shaping industrial policy and allocating resources for restructuring complicates the role of MITI in policy-formation and enforcement for any one industrial sector. The bureaucracy also commands comparatively smaller financial resources to assist in the restructuring, depriving the state of the necessary quiver of incentives for persuading reluctant capitalists to move in common directions. MITI deals best with firms based in specific industries and might find close cooperation on market predictions and adjustment programs more difficult with the diversified textile firms today. As the textile firms continue to diversify at home and invest abroad, direct ties with MITI's Textile Bureau will continue to fade. One might also note a lack of consensus on issues amenable to joint action by state and capital, but crises can quickly generate compromise and consensus if survival is at stake. Certainly there are problems areas in which the state can be of assistance, such as the rising volume of imported textiles, but MITI has thus far refused to intervene. The firms themselves have changed. The diversified textile giants no longer enjoy among themselves the same well-defined, common set of interests within a single industry. In the past, employers' associations such as the JSA or the JCFA provided an effective forum for a common program of adjustment among member "textile" producers; they remain important today in helping to balance supply and demand among member mills. The associations face a new challenge of cooperation among the diversified firms of today's industry, yet the remarkable record of compromise and common action at the associations across the decades of change in the postwar industry bodes well for further adaptation and collective action.

Also, the profile of labor as an interest group has changed, in large part because of its growing importance in corporatist strategies of change. Generous benefits for severance, and strong efforts by the firms to retain the skilled workers suggest greater recognition of the role of labor in production and adjustment in the firm. Increased labor productivity, government oversight of labor relations at the firms, and the labor shortage in Japan have contributed to this renewed appreciation at the firms for their workers. The growing prominence of the Rengo coalition in the Diet has spurred hopes for greater political influence for the labor federation. Such changes might suggest textile workers today would command a more independent voice in any additional corporatist negotiations for adjustment, if it were not for continued reductions in textile production at the mogul firms. Reductions shift workers out of textile manufacture into other lines of production, strengthening identification with the firm and eroding identification with a common industrial craft across firms. Labor today might be better informed on issues of offshore investment in textile production, or of diversification strategies because of the growth and diversification of the Zensen itself, but it still cannot assume a more autonomous role at the corporatist bargaining table.

Nonetheless, labor will continue to play a part in corporatist negotiations and will likely oppose more independent action by firms apart from corporatist patterns of change, and will certainly protest a return to bipartite bargaining between state and capital.

Collective Action

Analysis of adjustment and corporatist procedure answer our initial questions concerning direction and strategy. Adjustment tells us of structure, and corporatism of mediation. Study of adjustment turns our attention to result. Analysis of corporatist mediation sheds light on process. Conclusions about collective action draw us beyond process and consequence to consider causes and return to the question of why dissent did not shatter consensus among the moguls. The answer lies in the embedded and extended interests within and across firms in textiles that permitted a porous but effective solidarity in the face of decline. Capital, labor, and state found ways to extend immediate, special interests into long-range interests encompassing a combined, tripartite interest. Firms found common goals despite disparate interests and conflicting market choices, just as mill and merchant found cooperation congenial despite competing interests. Specialized roles of market coordination or fashion coordination, of cooperation at union, of compromise at association, and of authoritative coordination at MITI bring us close to cause, but so also do the encompassing interests that made the textile transition possible.

Direction could continue despite divergence because interests could be extended through channels such as industrial citizenship. Precedents of collective action offered fertile ground for corporatist strategies of mediation. Capital and labor hammered out compromises within the firm to permit mill closings and plant retooling. Executive and worker alike identified with the survival and success of the firm, and crises such as decline only highlighted the depth of such ties. The firm or enterprise commanded the center of the economy, rather than separate and competing groups of either capital or labor. That labor and capital within the firm could extend their separate, special interests into a more comprehensive advantage of the company provides a model for extension of interests in the world beyond the firm. That identification with the interests of something larger than either capital or labor alone has shaped negotiation across decades within the firm suggests roots and precedents not easily reversed.

There are multiple patterns of extension beyond the firm. Moguls and mavericks stretched their particular interests into an association interest in adjustment, despite divergence and dissent. State and capital put together a common compass of change, without denying the individual interests. Mills and trading houses and mills and fashion houses found ways to expand the separate interests of maker and merchant into profitable cooperation despite decline. Transnational ventures in offshore production represent one of the more dramatic cases of extending interests across decades and borders despite changing investment climates at home and abroad. Interests extended reinforce the long-term rooting of productive compromise, and such precedent provides fertile ground in turn for further melding of special and comprehensive goals. We have cited permeable boundaries between labor and management, just as Fruin highlighted fluid boundaries among firms apparent in interfirm cooperation, and Eccleston emphasized the weak distinction between state and society evident in Japan.[5] Encompassing interests stretch conventional borders within and across enterprises.

Extended and embedded interests, coupled with the institutions and routinized procedures fostering rearticulation of special goals into broader interests, gave character to the textile transition. Case studies of adjustment at the level of the firm and federation and association histories of restructuring will shed further light on cause and collective action. Scrutiny of embedded anomalies of the textile transition in ties between mill and merchant revealed reason in apparently inefficient market relations. Closer study of patterns of negotiation may well reveal other rationalities laced within opaque and apparently inefficient commercial practices. Corporatist strategies mediating and channeling patterns of collective action adds further light. A focus on corporatism within the larger context of collective behavior clarifies structure and procedures in extending interests. The dual emphasis on embedded and extended interests suggests a final insight significant for understanding collective behavior in a Japanese context. Attention to both context and process, to strategies of change, and to anomalies as well as regularities offers a promising direction for further study of adjustment.

5. Mark W. Fruin, *The Japanese Enterprise System: Competitive Strategies and Cooperative Structures* (Oxford: Clarendon Press, 1992); Bernard Eccleston, *State and Society in Post-War Japan* (Cambridge: Basil Blackwell and Polity Press, 1989).

Index